ROBUSTNESS IN STATISTICS

Academic Press Rapid Manuscript Reproduction

Proceedings of a Workshop
Sponsored by the Mathematics Division, Army Research Office
Held at Army Research Office, Weiss Building, April 11–12, 1978

ROBUSTNESS IN STATISTICS

edited by

ROBERT L. LAUNER

U.S. Army Research Office
Mathematics Division
Research Triangle Park, North Carolina

GRAHAM N. WILKINSON

Mathematics Research Center
University of Wisconsin–Madison
Madison, Wisconsin
and
Genetics Department
University of Adelaide
South Australia

ACADEMIC PRESS New York San Francisco London 1979
A Subsidiary of Harcourt Brace Jovanovich, Publishers

ACADEMIC PRESS, INC.
111 Fifth Avenue, New York, New York 10003

United Kingdom Edition published by
ACADEMIC PRESS, INC. (LONDON) LTD.
24/28 Oval Road, London NW1 7DX

Library of Congress Cataloging in Publication Data
Main entry under title:

Robustness in statistics.

1. Robust statistics—Congresses. I. Launer, Robert L. II. Wilkinson, Graham N. III. United States. Army Research Office, N.C. Mathematics Division.
QA276.A1R6 519.5 79-13893
0-12-438150-2

PRINTED IN THE UNITED STATES OF AMERICA

79 80 81 82 9 8 7 6 5 4 3 2 1

CONTENTS

CONTRIBUTORS

Numbers in parentheses indicate the pages on which authors' contributions begin.

William S. Agee (107), Mathematical Services Branch, Analysis and Computation Division, National Range Operations Directorate, US Army White Sands Missile Range, White Sands Missile Range, New Mexico 88002

D. F. Andrews (19), Department of Statistics, University of Toronto, Toronto, Ontario, Canada M5S 1A1

G. E. P. Box (201), Department of Statistics and Mathematics Research Center, University of Wisconsin—Madison, Madison, Wisconsin 53706

H. A. David (61), Statistical Laboratory, Iowa State University, Ames, Iowa 50011

Robert V. Hogg (1), Department of Statistics, The University of Iowa, Iowa City, Iowa 52242

Peter J. Huber (33), Swiss Federal Institute of Technology, Zürich, Switzerland

M. Vernon Johns (49), Department of Statistics, Stanford University, Stanford, California 94305

N. L. Johnson (127), Department of Statistics, University of North Carolina, Chapel Hill, North Carolina 27514

R. Douglas Martin (147), Departments of Electrical Engineering and Biostatistics, University of Washington, Seattle, Washington 98195; Bell Laboratories, Murray Hill, New Jersey 07947

Emanuel Parzen (237), Institute of Statistics, Texas A&M University, College Station, Texas 77843

John W. Tukey (75, 103), Department of Statistics, Princeton University, Princeton, New Jersey 08540

Robert H. Turner (107), Mathematical Services Branch, Analysis and Computation Division, National Range Operations Directorate, US Army

White Sands Missile Range, White Sands Missile Range, New Mexico 88002

V. David VandeLinde (177), Department of Electrical Engineering, Johns Hopkins University, Baltimore, Maryland 21218

Graham N. Wilkinson (259), Mathematics Research Center, University of Wisconsin—Madison, Madison, Wisconsin 53706 and Genetics Department, University of Adelaide, South Australia (present address)

PREFACE

This book contains the proceedings of a workshop on Robustness in Statistics held on April 11–12, 1978, at the Army Research Office, Research Triangle Park, North Carolina. The stated purpose of this workshop was to present a state-of-the-art review of "statistical methods which are relatively insensitive to: departure from distributional assumptions, outliers, sample censoring or other modifications, or large sample requirements." This is a very broad interpretation of statistical robustness, but it was used in the original invitations to encourage participation from a relatively broad cross section of statistical researchers. Invitations were given to twelve people, eight of whom are, or until recently were, supported by the ARO. The overall plan was to develop a program that would begin at an introductory level and progress logically through recent theoretical developments, special applications, and on to new approaches to robustness. This was also used as a general plan for the proceedings.

There are several features about this book of interest to the serious reader. The volume begins with an excellent introduction to statistical robustness by Robert Hogg. There is also a brief, second paper by John Tukey summarizing many of the important points made in the various presentations. This paper informally separates the "theory" and the "special applications" sections. Three members of the Princeton study group (David Andrews, Peter Huber, and John Tukey) and one other collaborator (Vernon Johns) are contributors to this volume. The format of the volume has a novel feature that was suggested by Graham Wilkinson—an introductory section containing short abstracts of each paper in the order in which they appear. This should be especially appealing to the busy reader. Finally, there are several new approaches to robustness by George Box (who introduced the word *robust* to the statistics vocabulary), Emanuel Parzen, and Graham Wilkinson. It is regrettable that the many lively discussions stimulated by the interesting talks could not be recorded in these pages. The perspicacity versus pertinence debate led by George Box and John Tukey was especially noteworthy.

The difficult task of chairing the sessions of the very compact agenda was competently handled by: Raymond J. Carroll, University of North Carolina, Chapel Hill, North Carolina; Charles Quesenberry, North Carolina State University, Raleigh, North Carolina; Donald Burdick, Duke University, Durham, North Carolina; Robert W. Turner, White Sands Missile Range, New Mexico.

Mrs. Sherry Duke and Mrs. Francis Atkins of the Army Research Office attended cheerfully to the many details of organizing and running the conference. The difficult job of organizing the papers and the glossary was most competently handled by Mrs. Sally C. Ross of the Mathematics Research Center. Finally, a special thank you to Bob Hogg for his ideas which helped make this a successful workshop.

Robert L. Launer

ABSTRACTS

AN INTRODUCTION TO ROBUST ESTIMATION

Robert V. Hogg

In many sets of data, there are fairly large percentages of "outliers" due to heavy tailed models or errors in collecting and recording. Since outliers have an unusually great influence on "least squares" estimators (or generalizations of them), robust procedures attempt to modify those schemes. One such method is M-estimation, and the terminology and associated techniques are explained: ρ and ψ functions, iteration schemes, error structure of estimators, sensitivity and influence curves, and their connection with jackknife estimates. Also L- and R-estimation and adaptive estimators are mentioned. A few examples are given and comments are made about the use of robust procedures today.

THE ROBUSTNESS OF RESIDUAL DISPLAYS

D. F. Andrews

The examination of least-squares residuals is a widely used procedure for constructing useful linear and nonlinear models. Practice has shown this to be a useful technique, even though it does not clearly indicate departures from distributional assumptions. This paper introduces a simple analog of trimmed means, useful for studying residuals from a robust point-of-view. In particular, this procedure clearly highlights departures from the assumed distribution.

ROBUST SMOOTHING

Peter J. Huber

A conceptually simple approach to robust smoothing of curve data is sketched, based on ideas gleaned from spline smoothing, robust regression, and time series analysis. If X_t is the original time series, Z_t the smoothed one, this approach minimizes ave $\{(Z''_t)^2\}$ subject to a side condition of the form ave $\{\rho(X_t - Z_t)\} \leq$ const., where ρ is a symmetric convex function with a bounded derivative.

ROBUST PITMAN-LIKE ESTIMATORS

M. Vernon Johns

A new class of robust estimators (called P-estimators) based on the location and scale invariant Pitman estimators of location is introduced. These estimators are obtained by replacing the probability density function appearing in the ordinary Pitman estimator by a suitable symmetric nonnegative function which need not be integrable. Scale invariance is obtained by the use of a scale functional related in a similar way to the Pitman estimator for scale. Influence functions and asymptotic variances are derived. Triefficiency comparisons, based on simulations, are made between P-estimators and the "bisquare" M-estimator for sample size twenty. Generalizations to the linear regression problem are briefly considered.

ROBUST ESTIMATION IN THE PRESENCE OF OUTLIERS

H. A. David

The bias and mean-square error of various scale estimators, expressible as linear functions of order statistics, are studied in a sample of size n. Specific attention is paid to the cases when the outlier comes from a population differing from the target population either in location or scale (slippage models). Exact numerical results for 12 estimators of the standard deviation are given for $n = 10$ in the case of slippage of the mean when the target population is normal. Also discussed are relations between (a) models for outliers and mixture models and (b) robust estimation and tests for outliers.

STUDY OF ROBUSTNESS BY SIMULATION: PARTICULARLY IMPROVEMENT BY ADJUSTMENT AND COMBINATION

John W. Tukey

Monte Carlo (or experimental sampling) studies of performance of statistics should be an active process of data analysis, learning, and improvement. Useful techniques for study and improvement can be formalized. Among those discussed are use of linear combinations, adjustment of level by compartment, and selection of estimates by compartment. All of these can be formalized and carried out in terms of the multiple criteria appropriate for robust resistant techniques. Both choices, in their use and results, are discussed. A variety of approaches to optimization according to multiple criteria are formulated and discussed, as are generalization to polyregression.

APPLICATION OF ROBUST REGRESSION TO TRAJECTORY DATA REDUCTION

William S. Agee and Robert H. Turner

Robust statistics provides a fresh approach to the difficult problem of editing in data reduction. Of prime concern are grossly erroneous measurements which, when undetected, completely destroy automated data reduction procedures causing costly reruns and time delays with human detection of the erroneous measurements. The application of robust statistical methods has been highly successful in dealing with this problem. An introduction to the robust M-estimates and their numerical computation is given. The applications of M-estimates to data preprocessing, instrument calibration, N-station cinetheodolites, N-station radar solution, and filtering are described in detail. Numerical examples of these applications using real measurements are given.

TEST FOR CENSORING EXTREME VALUES (ESPECIALLY) WHERE POPULATION DISTRIBUTIONS ARE INCOMPLETELY DEFINED

Norman L. Johnson

If a set of univariate observations is given, which may be the remainder of a random sample after removal of specified extreme ordered values, a test of the hypothesis that such censoring has not occurred cannot be constructed in the absence of all knowledge of the population distribution of the measured character. When this knowledge is complete, exact tests can be constructed. The present paper discusses construction of tests when knowledge is only partial, ranging from knowledge of the form of population distribution (e.g., normal, exponential) to availability of a known complete random sample from the population, to knowledge that some one among several sets of values corresponds to a complete random sample.

ROBUST ESTIMATION FOR TIME SERIES AUTOREGRESSIONS

R. Douglas Martin

This paper summarizes the status of research on robust estimation of the parameters of time series autoregressions. Two distinct types of outliers are considered, namely innovations outliers (IO) and additive observational errors outliers (AO). The lack of robustness of the least-squares estimates toward both types of outliers is discussed. It is shown that M-estimates of autoregressive parameters, including the mean, are efficiency robust toward IO situations. However, M-estimates are not robust toward AO situations; so two alternative classes of robust estimates are presented: (i) generalized M-estimates and (ii) conditional-mean M-estimates. After discussing these estimates, two methods for assessing outlier type are described.

ROBUST TECHNIQUES IN COMMUNICATION

V. David VandeLinde

Three problems arising in communication theory are considered from the viewpoint of minmax robustness. The areas investigated are estimation, detection, and quantization of a stochastic source. Some general results are presented for the Robbins-Monro stochastic algorithm as a robust location estimate and in extending the minmax estimates of location to larger distribution sets. General approaches for the design of minmax robust detectors are presented for both fixed sample and sequential detection problems. Finally, robust quantization is considered for unimodally distributed input signals satisfying a generalized moment constraint.

ROBUSTNESS IN THE STRATEGY OF SCIENTIFIC MODEL BUILDING

G. E. P. Box

The philosophy of robust procedures is discussed. It is argued that the present emphasis by statistical researchers on ad hoc methods of robust estimation is mistaken. Classical methods of estimation should be retained using models that more appropriately represent reality. Attention should not be confined merely to discrepancies arising from outliers and heavy tailed distributions but should be extended to include serial dependence, need for transformation, and other problems. Some researches of this kind using Bayes' theorem are discussed.

A DENSITY–QUANTILE FUNCTION PERSPECTIVE ON ROBUST ESTIMATION

Emanuel Parzen

A perspective on robust estimation is discussed, with three broad sets of conclusions. Point I: The means must be justified by the ends. Point II: Graphical goodness-of-fit procedures should be applied to data to check whether they are adequately fitted by the qualitatively defined models that are implicitly assumed by robust estimation procedures. Point III: There is a danger that researchers may regard robust regression procedures as a routine solution to the problem of modeling relations between variables without first studying the distribution of each variable. New tools introduced include: Student's window; quantile–Box plots; density–quantile estimation approach to goodness-of-fit tests; and a definition of statistics as "arithmetic done by the method of Lebesgue integration."

ROBUST INFERENCE—THE FISHERIAN APPROACH

Graham N. Wilkinson

The Fisherian theory of conditional parametric inference may well prove to be the soundest theoretical basis for a properly adaptive approach to statistical inference. However, some current robustness criteria are not compatible with the principles of this theory. The paper briefly describes the Fisherian theory and the conditional fiducial-confidence distributions to which it leads. A result of Hartigan is then given for producing "distribution-free" confidence distributions of location by applying an estimator to all possible subsamples of a given sample. Numerical comparisons of Fisher conditional and Hartigan confidence intervals favor the conditional approach, and point to new concepts of conditional central limit theory.

An Introduction to Robust Estimation

Robert V. Hogg

1. INTRODUCTION.

Certainly the method of least squares and generalizations of it have served us well for many years. However, it is recognized that "outliers," which arise from heavy tailed distributions or are simply bad data points due to errors, have an unusually large influence on the least squares estimators. That is, the outliers pull the least squares "fit" towards them too much; a resulting examination of the residuals is misleading because then they look more like normal ones. Accordingly, robust methods have been created to modify least squares schemes so that the outliers have much less influence on the final estimates. One of the most satisfying robust procedures is that given by a modification of the principle of maximum likelihood; hence we begin with that approach. First, however, it should be mentioned that a new paperback [8] by Huber provides an excellent summary of many of the mathematical aspects of robustness.

2. M-ESTIMATION.

Let $X_1, X_2, \cdots, X_n$ be a random sample that arises from a distribution with density $f(x-\theta)$ of the continuous type, where θ is a location parameter. The logarithm of the likelihood function is

$$\ln L(\theta) = \sum_{i=1}^{n} \ln f(x_i-\theta) = -\sum_{i=1}^{n} \rho(x_i-\theta),$$

ISBN 0-12-438150-2

where $\rho(x) = -\ln f(x)$. In maximum likelihood we wish to maximize $\ln L(\theta)$ or, in terms of the ρ function, minimize $K(\theta) = \Sigma\rho(x_i - \theta)$. Suppose that this minimization can be achieved by differentiating and solving $K'(\theta) = 0$; that is, finding the appropriate θ that satisfies

$$\sum_{i=1}^{n} \psi(x_i - \theta) = 0,$$

where $\psi(x) = \rho'(x) = -f'(x)/f(x)$. The solution of this equation that minimizes $K(\theta)$ is called the maximum-likelihood or M-estimator of θ and is denoted by $\hat{\theta}$.

Three typical classroom examples of this process are given by the following distributions, the first of which provides the least squares estimate and the second, the "least absolute values" estimate.

(1) Normal: $\rho(x) = \frac{x^2}{2} + c$, $\psi(x) = x$,

$$\sum_{i=1}^{n} (x_i - \theta) = 0 \quad \text{yields} \quad \hat{\theta} = \bar{x}.$$

(2) Double exponential: $\rho(x) = |x| + c$, $\psi(x) = \begin{cases} -1, & x < 0, \\ 1, & x > 0, \end{cases}$

$$\sum_{i=1}^{n} \psi(x_i - \theta) = 0 \quad \text{yields} \quad \hat{\theta} = \text{sample median}.$$

(3) Cauchy: $\rho(x) = \ln(1+x^2) + c$, $\psi(x) = \frac{2x}{1+x^2}$,

$$\sum_{i=1}^{n} \psi(x_i - \theta) = 0 \quad \text{is solved by numerical methods.}$$

It is interesting to note that the ψ functions of examples (2) and (3) are bounded and that of (3) even redescends and approaches zero asymptotically. We note that the solutions in (2) and (3) are not influenced much by outliers and thus are robust. On the other hand, the least squares estimate $\bar{x}$ of (1) is not robust since extreme values do greatly influence $\bar{x}$.

Thus in robust M-estimation we wish to determine a ψ function so that the resulting estimator will protect us against a small percentage (say, around 10%) of outliers. But, in addition, we desire these procedures to produce reasonably good (say, 95% efficient) estimators in case the data actually enjoys the normal assumptions. Here the 5%

efficiency that we sacrifice in these cases is sometimes referred to as the "premium" that we pay to gain all that protection in very non-normal cases.

Using a more technical definition of robustness, Huber [7] derived the following robust ρ and ψ.

$$\rho(x) = \begin{cases} \frac{x^2}{2}, & |x| \le a \\ a|x| - \frac{a^2}{2}, & |x| > a \end{cases} \quad ; \qquad \psi(x) = \begin{cases} -a, & x < -a \\ x, & |x| \le a \\ a, & x > a. \end{cases}$$

These are the ρ and ψ functions associated with a distribution which is "normal" in the middle with "double exponential" tails. The corresponding M-estimator is actually the minimax solution

$$\min_T \max_F [\text{asym. var. } (T)],$$

where F is a family of ϵ-contaminated normal distributions $\{(1-\epsilon)\Phi + \epsilon H\}$ and T varies over a class of estimators. See Huber's 1964 article [7] for the details of this development.

There is one difficulty associated with Huber's ψ (also with many other ψ functions too): For example, if the x values are multiplied by a constant, the original estimator is not necessarily multiplied by that same constant. To create a scale invariant version of the M-estimator, we find the solution $\hat{\theta}$ of

$$\sum_{i=1}^{n} \psi\left(\frac{x_i - \theta}{s}\right) = 0$$

where s is a robust estimate of scale such as

$$s_1 = \frac{\text{median}|x_i - \text{median } x_i|}{.6745}$$

or

$$s_2 = \frac{75^{th} \text{ percentile} - 25^{th} \text{ percentile}}{2(.6745)}.$$

If the sample arises from a normal distribution, then s_1 and s_2 are estimates of σ.

With Huber's ψ and one of these robust estimates of scale, we would take the "tuning constant" a to be about

1.5. The reason for this selection is that if the sample actually comes from a normal distribution, most of the items would enjoy the property that $|x_i-\theta|/s \le 1.5$. If all satisfied this inequality, then $\psi[(x_i-\theta)/s] = (x_i-\theta)/s$ and $\Sigma\psi[(x_i-\theta)/s] = 0$ has the solution $\hat{\theta} = \bar{x}$, as is desired in the normal case. It has been established [6] that if σ is known, then $a = 1.345$ is needed to achieve asymptotic efficiency of 95% (or a premium of 5%) under normal assumptions.

Let us mention three iteration schemes that can be used to solve $\Sigma\psi[(x_i-\theta)/s] = 0$, where ψ is that of Huber. We use $\tilde{\theta}_1$ = sample median as the first guess. Of course, $\tilde{\theta}_2$ is used as the second guess in the iteration, and so on.

Newton's

$$\tilde{\theta}_2 = \tilde{\theta}_1 + \frac{s\Sigma\psi[(x_i-\tilde{\theta}_1)/s]}{\Sigma\psi'[(x_i-\tilde{\theta}_1)/s]},$$

where the denominator of the second term counts the number of items that enjoy $|x_i-\tilde{\theta}_1|/s \le a$.

H-Method

$$\tilde{\theta}_2 = \tilde{\theta}_1 + \frac{s\Sigma\psi[(x_i-\tilde{\theta}_1)/s]}{n},$$

where the second term is the average of the pseudo (Winsorized) residuals (that is, "least squares" is applied to the pseudo residuals).

Weighted Least Squares

$$\tilde{\theta}_2 = \frac{\Sigma w_i x_i}{\Sigma w_i} = \tilde{\theta}_1 + \frac{s\Sigma\psi[(x_i-\tilde{\theta}_1)/s]}{\Sigma w_i},$$

where the weight is

$$w_i = \frac{\psi[(x_i-\tilde{\theta}_1)/s]}{(x_i-\tilde{\theta}_1)/s}, \quad i = 1,2,\cdots,n.$$

Note that these expressions are very similar. With Huber's ψ, $n \ge \Sigma w_i \ge \Sigma\psi'$; thus Newton's method provides the greatest adjustment on the first iteration. In fact, using Newton's, if two adjacent iterations have the same items in the "wings" and central part, then the solution has been reached and no more iterations are needed.

Other ψ functions that are commonly used are the following, with suggested values of the tuning constants [6].

Hampel's ψ

$$\psi(x) = (\text{sign } x)\begin{cases} |x| & , \quad 0 \le |x| < a, \\ a & , \quad a \le |x| < b, \\ \dfrac{c-|x|}{c-b} & , \quad b \le |x| < c, \\ 0 & , \quad c \le |x|. \end{cases}$$

Reasonably good values are $\underline{a} = 1.7$, $\underline{b} = 3.4$, $\underline{c} = 8.5$.

Andrew's sine

$$\psi(x) = \begin{cases} \sin(x/a) & , \quad |x| \le a\pi \\ 0 & , \quad |x| > a\pi \end{cases}$$

with $\underline{a} = 2.1$. Actually if the scale is known, $\underline{a} = 1.339$ requires a premium of 5%.

Tukey's biweight

$$\psi(x) = \begin{cases} x[1-(x/a)^2]^2 & , \quad |x| \le a, \\ 0 & , \quad |x| > a, \end{cases}$$

with $\underline{a} = 6.0$. If the scale is known, $\underline{a} = 4.685$ means the premium is 5%.

Since the ρ functions associated with these redescending ψ functions is not convex, there could be certain convergence problems in the iteration procedures. In any case, Collins [2] showed that the ψ should not redescend "too steeply."

3. REGRESSION.

First, consider the linear model

$$\underset{\sim}{Y} = \underset{\sim}{X}\underset{\sim}{\beta} + \underset{\sim}{\varepsilon},$$

where the design matrix $\underset{\sim}{X}$ is nxp, the parameter vector $\underset{\sim}{\beta}$ is px1, and the random nx1 vector $\underset{\sim}{\varepsilon}$ has mean vector $\underset{\sim}{0}$

and covariance matrix $\sigma^2 \underset{\sim}{I}$. To obtain a robust estimate of σ and a start for the iteration process, we need a preliminary estimate of $\underset{\sim}{\beta}$. It is helpful if this is robust and frequently the "least absolute value" estimate is used. In any case, we denote this by $\hat{\underset{\sim}{\beta}}_1$. A robust measure of spread is

$$s = \underset{\substack{\text{(of the nonzero)} \\ \text{deviations}}}{\text{median}} \frac{|y_i - \underset{\sim}{x}_i \hat{\underset{\sim}{\beta}}_1|}{.6745},$$

where $\underset{\sim}{x}_i$ is the ith row of the design matrix $\underset{\sim}{X}$. Here we take the median of the nonzero deviations because, with large p, too many residuals can equal zero and in those cases the usual median yields an s that is too small and accordingly a premium that is too large.

We are required to

$$\text{minimize} \sum_{i=1}^{n} \rho\left(\frac{y_i - \underset{\sim}{x}_i \underset{\sim}{\beta}}{s}\right).$$

Considering the p first partial derivatives, this means we must solve the p equations

$$\sum_{i=1}^{n} \psi\left(\frac{y_i - \underset{\sim}{x}_i \underset{\sim}{\beta}}{s}\right) x_{ij} = 0, \quad j = 1,2,\cdots,p,$$

simultaneously. Again we could use iterations based upon Newton's method, the H-method, or weighted least squares. While the H-method has some advantage in the linear model (only the inverse of $\underset{\sim}{X}'\underset{\sim}{X}$ is needed in the least squares solutions using the pseudo residuals), weighted least squares is described as it is more general (linear or nonlinear models).

We replace the p equations by the approximations

$$\sum_{i=1}^{n} w_i x_{ij} (y_i - \underset{\sim}{x}_i \underset{\sim}{\beta}) \approx 0, \quad j = 1,2,\cdots,p,$$

where

$$w_i = \frac{\psi[(y_i - \underset{\sim}{x}_i \hat{\underset{\sim}{\beta}}_1)/s]}{(y_i - \underset{\sim}{x}_i \hat{\underset{\sim}{\beta}}_1)/s}, \quad i = 1,2,\cdots,n.$$

The solution to these approximate equations is

$$\hat{\beta}_2 = (X'WX)^{-1}X'WY,$$

where W is the diagonal matrix $\mathrm{diag}(w_1,\cdots,w_n)$. Of course, $\hat{\beta}_2$ provides us with a new start and new weights, which requires recomputation of $(X'WX)^{-1}$. This iteration is continued until a reasonable degree of convergence is reached. The solution is denoted by $\hat{\beta}$ and the final weight given to an item will indicate whether or not it is an outlier.

While there is not complete accord as to the error structure of $\hat{\beta}$, asymptotic theory suggests that $\sqrt{n}(\hat{\beta}-\beta)$ has an approximate p-variate normal distribution with mean vector 0 and covariance matrix

$$\left(\frac{n}{n-p}\right)s^2 \frac{\frac{1}{n}\Sigma\psi^2[(y_i - x_i\beta)/s]}{\{\frac{1}{n}\Sigma\psi'[(y_i - x_i\beta)/s]\}^2} (X'X)^{-1}.$$

In our discussions thus far, no mention has been made of iterating on s also. While most robustniks do iterate on s also, there are some problems of convergence created. Clearly more research is needed on scale problems in general, and on the error structure of $\hat{\beta}$ in particular.

Let us now consider the case in which $x_i\beta$ is replaced by the nonlinear function $h(x_i,\beta)$. In order to

$$\text{minimize} \sum_{i=1}^{n} \rho\left[\frac{y_i - h(x_i,\beta)}{s}\right],$$

with an appropriate robust estimator s of scale, we solve the p equations

$$\sum_{i=1}^{n} \psi\left[\frac{y_i - h(x_i,\beta)}{s}\right] \frac{\partial h(x_i,\beta)}{\partial \beta_j} = 0,$$

$j = 1,2,\cdots,p$, simultaneously. This can be done by iteration using weighted least squares, with weights

$$w_i = \frac{\psi[(y_i - h(x_i,\hat{\beta}_1))/s]}{[y_i - h(x_i,\hat{\beta}_1)]/s}, \quad i = 1,2,\cdots,n,$$

on the first iteration, where β_1 is a preliminary estimate of β.

4. RIDGE REGRESSION.

In the linear model, mention should be made how to robustify ridge regression (or possibly Bayesian estimates in general). With the usual normal assumptions, $\hat{\beta} = (X'X)^{-1}XY$ has a p-variate normal distribution with parameters β, $\sigma^2(X'X)^{-1}$. If β has a prior p-variate normal distribution with parameters β_0, Σ_0, it is well known that the post mean of β, given $\hat{\beta}$, is

$$\left(\frac{1}{\sigma^2} X'X + \Sigma_0^{-1}\right)^{-1}\left(\frac{1}{\sigma^2} X'Y + \Sigma_0^{-1}\beta_0\right).$$

If we "shrink" to $\beta_0 = 0$ and take $\Sigma_0 = \sigma_0^2 I$, we obtain the usual ridge estimator

$$(X'X + kI)^{-1}X'Y, \quad \text{where} \quad k = \sigma^2/\sigma_0^2.$$

Clearly, there is no need to shrink each element of the estimator towards zero if other prior ideas are held and the prior covariance matrix of β could be $\text{diag}(k_1, k_2, \cdots, k_p)$ or even one with nonzero covariances. In the general or the special case, respectively, let us augment the design and observation matrices with p additional rows to obtain

$$X_A = \begin{pmatrix} X \\ \sigma\Sigma_0^{-1/2} \end{pmatrix} = \begin{pmatrix} X \\ \sqrt{k}\ I \end{pmatrix}; \quad Y_A = \begin{pmatrix} Y \\ \sigma\Sigma_0^{-1/2}\beta_0 \end{pmatrix} = \begin{pmatrix} Y \\ 0 \end{pmatrix}.$$

Then the usual least squares estimator with these $n+p$ observations (p of which are pseudo) provides the ridge estimators

$$\begin{aligned}(X_A'X_A)^{-1}X_A'Y &= (X'X + \sigma^2\Sigma_0^{-1})^{-1}(X'Y + \sigma^2\Sigma_0^{-1}\beta_0) \\ &= (X'X + kI)^{-1}(X'Y),\end{aligned}$$

in the general and special case, respectively. Now if we replace "least squares" on X_A, Y_A by a robust scheme, some of the $n+p$ observations could have low weights, indicating that they might be outliers. Among these outliers, there might be a few of the p pseudo observations which could suggest that those corresponding prior assignments simply do not agree with the real observations. Of course these priors would automatically be given low or zero weight by the robust procedure in the final estimator. It seems as if ridge estimator, with this robust look, would be most useful in applications.

5. L-ESTIMATION.

Let us first consider the simple case of a random sample from a distribution of the continuous type that has a location parameter θ. Say the order statistics of the sample are $X_{(1)} \leq X_{(2)} \leq \cdots \leq X_{(n)}$. An L-estimator is one which is a linear combination of these order statistics. Examples of L-estimators are:

(a) sample median,

(b) α-trimmed mean $\bar{X}_\alpha = \frac{1}{n-2[n\alpha]} \sum_i X_{(i)}$, where the summation is over $i = [n\alpha]+1, \cdots, n-[n\alpha]$,

(c) Gastwirth's estimator which is a weighted average of the $33\frac{1}{3}$rd, 50th, and $66\frac{2}{3}$rd percentiles with respective weights .3, .4, and .3,

(d) Tukey's trimean which is a weighted average of the 1st, 2nd, and 3rd quartiles with respective weights $\frac{1}{4}$, $\frac{1}{2}$, and $\frac{1}{4}$.

The generalization of L-estimators to the regression situation is not as clear as in the case of M-estimators. However, since the use of the ρ function, $\rho(x) = |x|$, yields the median as an estimator (and the "median plane or surface" in the regression situation), this could easily be modified to get other percentiles. That is, the ρ function

$$\rho(x) = \begin{cases} -(1-p)x, & x < 0, \\ px, & x \geq 0, \end{cases}$$

yields the $(100p)^{th}$ percentile in the single sample case and thus estimates of the "$(100p)^{th}$ percentile plane or surface" in the regression situation. Clearly, generalizations of estimates like those of Gastwirth or Tukey could now be constructed in regression problems. Moreover, it seems as if in many situations (say like educational data involving prediction of college performance from high school rank and

SAT or ACT scores) we would be interested in estimates of some percentiles other than those of the middle. Thus percentile estimates could stand on their own as well as in combination with others to predict a "middle" plane or surface.

6. R-ESTIMATION.

R-estimation is a nonparametric method resulting from ranking when the sample arises from a continuous-type distribution. It can easily be extended to regression. Consider the linear model and modify least squares by replacing one factor in $(y_i - \underset{\sim}{x}_i \underset{\sim}{\beta})^2$ by the rank of $y_i - \underset{\sim}{x}_i \underset{\sim}{\beta}$, say R_i. The rank R_i is clearly a function of $\underset{\sim}{\beta}$. Hence we wish to

$$\text{minimize } \sum_{i=1}^{n} (y_i - \underset{\sim}{x}_i \underset{\sim}{\beta}) R_i .$$

This, in turn, can be generalized by replacing the ranks $1, 2, \cdots, n$ by the "scores"

$$a(1) \le a(2) \le \cdots \le a(n).$$

Thus, in this generalized setting, we wish to

$$\text{minimize } \sum_{i=1}^{n} (y_i - \underset{\sim}{x}_i \underset{\sim}{\beta}) a(R_i).$$

Of course, two examples of scores are

(a) Wilcoxon scores: $a(i) = i$ or ranks,

and

(b) Median scores: $a(i) = \begin{cases} -1 , & i < (n+1)/2, \\ 1 , & i > (n+1)/2. \end{cases}$

In 1972, Jaeckel [9] proved that this minimization is equivalent to solving the p-equations

$$\sum_{i=1}^{n} x_{ij} a(R_i) = 0 , \quad j = 1, 2, \cdots, p,$$

that must be solved approximately due to the discontinuities in $a(\cdot)$ and R_i. Of course, it is well known that "good" (certain asymptotic properties) scores are those given by

$$a(i) = \varphi\left(\frac{i}{n+1}\right), \quad \text{where} \quad \varphi(t) = -\frac{f'[F^{-1}(t)]}{f[F^{-1}(t)]} .$$

Examples of this are:

(a) f normal produces $\varphi(t) = \Phi^{-1}(t)$, $0 < t < 1$, that gives normal scores;

(b) f double exponential produces

$$\varphi(t) = \begin{cases} -1, & 0 < t < \frac{1}{2} \\ 1, & \frac{1}{2} < t < 1 \end{cases}$$

that gives median scores;

(c) f logistic produces $\varphi(t) = 2t-1$, $0 < t < 1$, that gives Wilcoxon scores.

In 1977, Jurečková [10] proved that, with certain scores $a(\cdot)$ and ψ functions, the R-estimators and M-estimators are asymptotically equivalent. Among other conditions, we need that

$$\varphi(t) = c_1 \psi[F^{-1}(t)] + c_2,$$

where c_1 and c_2 are constants, for this equivalence.

7. SENSITIVITY AND INFLUENCE CURVES AND THE JACKKNIFE.

Let us consider a special case of a sensitivity curve. If $n = 5$, the trimmed mean $\bar{x}_{\alpha,n}$, with $\alpha = 1/5$, would be the average of the remaining items after trimming the smallest and largest items. If an extra x value is introduced and the difference $\bar{x}_{\alpha,n+1} - \bar{x}_{\alpha,n}$ is considered as a function of x, it is called a sensitivity curve. Here it is, with $\bar{x}_{\alpha,n} = (x_{(2)} + x_{(3)} + x_{(4)})/3$,

$$\bar{x}_{\alpha,n+1} - \bar{x}_{\alpha,n} = \begin{cases} \dfrac{x_{(1)}+x_{(2)}+x_{(3)}+x_{(4)}}{4} - \bar{x}_{\alpha,n}, & x < x_{(1)}, \\ \dfrac{x + x_{(2)}+x_{(3)}+x_{(4)}}{4} - \bar{x}_{\alpha,n}, & x_{(1)} \le x \le x_{(5)}, \\ \dfrac{x_{(2)}+x_{(3)}+x_{(4)}+x_{(5)}}{4} - \bar{x}_{\alpha,n}, & x_{(5)} < x. \end{cases}$$

That is, $\bar{x}_{\alpha,n+1} - \bar{x}_{\alpha,n}$ is continuous and is a line segment with slope $1/4$ when $x_{(1)} \le x \le x_{(5)}$ and is equal to constants when $x < x_{(1)}$ and $x_{(5)} < x$. Note that this sensitivity curve is bounded, but the sensitivity curve for $\bar{x}$ would not be bounded.

The influence curve is like an asymptotic sensitivity curve (divided by $1/n+1$). Let us first consider an ideal estimator as a functional; for example, the "ideal $\bar{x}$" is the functional

$$T(F) = \int_{-\infty}^{\infty} x dF(x) = \int_0^1 F^{-1}(t)dt = \mu.$$

The influence curve for any functional $T(F)$ is the "derivative," where δ_x places probability one on x,

$$IC(x) = \lim_{\epsilon \to 0+} \frac{T[(1-\epsilon)F(x) + \epsilon\delta_x] - T[F(x)]}{\epsilon}.$$

For example, the influence curve of the trimmed mean is very similar to the sensitivity curve of $\bar{x}_{\alpha,n}$: constant, then a sloping line segment, then constant again. Under F, we have

$$\text{asym. var.}(\sqrt{n}\ T) = \int_{-\infty}^{\infty} [IC(x)]^2 dF(x)$$

and for M-estimators,

$$IC(x) = \frac{\psi(x)}{\int_{-\infty}^{\infty} \psi'(x)dF(x)} \propto \psi(x).$$

Clearly, if $F(x)$ is a member of a family of distributions (some with long tails), we see that we want the $IC(x)$ to be a bounded function to keep the variance of $\sqrt{n}\ T$ reasonably small. For a more detailed discussion of these points, see the fine article by Hampel [4].

Let $T_n(X_1, X_2, \cdots, X_n)$ be an estimator of θ and say

$$E(T_n) = \theta + \frac{a_1}{n} + \frac{a_2}{n^2} + \cdots.$$

The pseudo values are defined by

$$T^*_{ni} = nT_n - (n-1)T_{n-1}(\text{with } X_i \text{ deleted}),$$

$i = 1,2,\cdots,n$. The jackknife is then the average of the pseudo values, namely,

$$T_n^* = \frac{1}{n}\sum_{i=1}^{n} T_{ni}^*.$$

Since

$$E(T_n^*) = \theta - \frac{a_2}{n(n-1)} + \cdots$$

does not contain the term involving $1/n$, the jackknife estimator reduces the bias. For example, the jackknife of the sample variance $s^2 = \Sigma(x_i-\bar{x})^2/n$ is the unbiased estimator $\Sigma(x_i-\bar{x})^2/(n-1)$ of σ^2.

Now the differences $T_{ni}^* - T_n^*$, $i = 1,2,\cdots,n$, form a collection that is almost the finite version of the influence curve and hence

$$\mathrm{var}(T_n) \approx \frac{\Sigma(T_{ni}^* - T_n^*)^2}{n(n-1)},$$

since this is an approximation of $E\{[IC(x)]^2/n\}$. Thus the jackknife and associated pseudo values provide an estimator of the variance of T_n; this is probably more valuable than the bias reducing property of the jackknife.

8. ADAPTIVE ESTIMATORS.

Only brief note is made here of adaptive estimators; but, if interested in more, the reader is referred to an expository article on the subject by Hogg [5]. The basic idea of adapting is the selection of the estimation procedure after observing the data. Thus, for example, the tuning constants or the amounts of trimming could be dictated by the sample. As a matter of fact, even the forms of the function $\psi(\cdot)$ and the score function $a(\cdot)$ could be selected after observing the sample. Of course, asymptotically, we can select the "best" $\psi(\cdot)$ or $a(\cdot)$, but most of the time we are working with sample sizes like 20, 30, or 50, not infinity. Hence we must find some reasonable procedures for those very limited sample sizes. One such scheme is to select a small class of underlying distributions that span a large collection of possible distributions. Then determine a good procedure for each member of that class. Finally let the observations select (analysis of residuals, plots, etc.) which procedure will actually be used by taking that distribution seemingly closest to what was observed. Incidentally, there is no objection to analyzing with all of the procedures: If they say the same thing, that is the answer; but if they differ, then the selection procedure is critical and it is most important that it be done well.

There has been some evidence [5,12] that adaptive procedures are of value. After all, if the "hubers, hampels,···" are good, wouldn't adaptive ones be better (particularly since the former are included in the latter)? Moreover, it seems as if the applied statisticians would find adaptation very appealing (they do it all the time anyway), and this gives us a chance to bring theory and applications closer together.

9. EXAMPLES.

(a) Andrews [1] reports on a set of data that had been analyzed in the text by Daniel and Wood [3]. There were 21 observational points, each with a response variable y and 3 independent variables x_1, x_2, x_3. The model

$$E(Y) = \beta_0 + \beta_1 x_1 + \beta_2 x_2 + \beta_3 x_3$$

was used on all 21 points and the β's were determined by least squares (Method A). However, by some astute analysis, they were able to discard 4 points as being outliers and least squares was used on the remaining 17 points (Method B). Recently, the same data with all 21 points has been analyzed by two robust procedures: Andrews' ψ function with $a = 1.5$ (Method C) and an R-procedure with median scores (Method D).

	Estimates		
Method	β_1	β_2	β_3
A	0.72	1.30	-0.15
B	0.80	0.58	-0.07
C	0.82	0.52	-0.07
D	0.83	0.58	-0.06

It is worthy of note that the two robust methods with all 21 points provide essentially the same estimates as does least squares with the 4 bad points set aside. Moreover, the robust methods found these 4 points automatically without having the insight of Daniel and Wood.

(b) Zeigler and Ferris [13] report that six laboratories initiated a sample exchange program to determine, among other things, the half-life of plutonium-241 (^{241}Pu). The ratio (say Y) of the contents of ^{241}Pu to that of ^{239}Pu was reported by each of the six labs every three months. This was continued until there were over 70 data points (note that 6 points were obtained each 3 months). The problem was to fit the nonlinear function $E(Y) = \beta_0 e^{-\beta_1 x}$, where x is the time. Along with the nonlinear least squares fit, a robust scheme using Andrews' ψ function was used. The results of the two procedures did differ. Moreover, the most interesting part of the analysis was that the final weights of 6 points were equal to zero, and all 6 of these were reported by the same lab! Based on this robust analysis, an investigation was made and it was found that this particular lab was making a technical error causing higher readings. This mistake was then corrected for future readings.

(c) Lenth [11] considered 51 simulated observations from a Cauchy distribution such that the median of each was on the curve $\sin(2\pi e^{-x^2})$, where the 51 x-values ranged from zero to 2.5. The conventional least-squares spline was fitted with six knots (x=0, .3, .7, 1.2, 1.8, 2.5). This was compared to two robust spline fits: Huber's ψ with $a = 1.1$ and Andrews' ψ with $a = 1.2$. The latter two reproduced the curve $\sin(2\pi e^{-x^2})$ much better than the least square fit. The results with Andrews' ψ were actually a little better than those of Huber's because here (with a Cauchy distribution) a redescending ψ is more appropriate.

10. THE USE OF ROBUSTNESS TODAY

A good applied statistician has always been on guard for outliers, discarding them or investing them more as is appropriate in a particular situation. However, in complicated data sets, it is most difficult to find some of these extreme points and a robust procedure can help in this regard. Hence we recommend that, in applications today, the following steps be carried out.

(a) Perform the usual (probably least squares) analysis.

(b) Also use a robust procedure, at least a one-step iteration. Ideally, Huber's ψ with $a = 1.5$ could be used for several iterations, followed by two or three steps with Andrews' ψ with $a = 1.5$. (Possibly from a preliminary investigation these tuning constants would be changed—that is, there would be some adapting.)

(c) If the results from (a) and (b) agree, report the usual statistical summaries associated with (a).

(d) If there is disagreement between (a) and (b), take another hard look at the data. In particular look at those points having low weights. Then questions like, "Has someone made a simple recording error?" to, "Is this outlier trying to tell us something?" can be asked.

These robust methods have been used successfully in many applications. For example, the Los Alamos Scientific Lab now has the option of using a robust procedure in each regression problem, and there they frequently exercise that option. Hopefully, in a few years, most statistical programs will have these options and applied statisticians will avail themselves of them if they have reason to be concerned about outliers.

REFERENCES

1. Andrews, D. F. (1974). A robust method for multiple linear regression. Technometrics, 16, 523-531.

2. Collins, J. R. (1976). Robust estimation of a location parameter in the presence of asymmetry. Ann. Statist., 4, 68-85.

3. Daniel, C. and Wood, F. S. (1971). Fitting Equations to Data, Wiley-Interscience, New York.

4. Hampel, F. R. (1974). The influence curve and its role in robust estimation. J. Amer. Statist. Assoc., 69, 383-393.

5. Hogg, R. V. (1974). Adaptive robust estimation. J. Amer. Statist. Assoc., 69, 909-927.

6. Holland, P. W. and Welsch, R. E. (1977). Robust regression using iteratively reweighted least squares. Commun. Statist., A6, 813-828.

7. Huber, P. J. (1964). Robust estimation of a location parameter. Ann. Math. Statist., 35, 73-101.

8. Huber, P. J. (1977). Robust Statistical Procedures, Society for Industrial and Applied Mathematics, Philadelphia.

9. Jaeckel, L. A. (1972). Estimating regression coefficients by minimizing the dispersion of the residuals. Ann. Math. Statist., 43, 1449-1458.

10. Jurečková, J. (1977). Asymptotic relations of M-estimates and R-estimates in linear regression models. Ann. Statist., 5, 464-472.

11. Lenth, R. V. (1977). Robust splines. Commun. Statist., A6, 847-854.

12. Wegman, E. J. and Carroll, R. J. (1977). A Monte Carlo study of robust estimators of location. Commun. Statist., A6, 795-812.

13. Zeigler, R. K. and Ferris, Y. (1973). Half-life of plutonium-241. J. Inorg. Nucl. Chem., 35, 3417-3418.

The author was partially supported by National Institute of Health Grant GM 22271-02.

Department of Statistics
The University of Iowa
Iowa City, Iowa 52242

The Robustness of Residual Displays

D. F. Andrews

1. Introduction

Linear models represent a large component of statistical analysis both directly and as approximations to non-linear models. These models were the first widespread applications of the current growth of robust estimation procedures. Robust regression procedures and robust analogues of analysis of variance are being implemented and used in many fields.

The development and theory of these robust procedures has centred around efficiency and resistance, both properties of the parameter estimates. However much of statistical practice deals with large data sets, with problems where high efficiency is of little practical importance but where the formulation of useful models is critical. Mallows (1977) presented several examples of this nature.

The formulation of useful models is an iterative process guided more by the structure of residuals than by the values of parameters. The structure of residuals is commonly assessed from graphical displays, plots of residuals against other variables or quantiles of some reference distribution. Much is known about the properties of least-squares residuals in such displays (suggesting that crude displays are more

ISBN 0-12-438150-2

complicated than some current practice would indicate.) What can be said about displays of residuals from robustly estimated models?

This paper presents examples of the problems associated with least-squares residuals and investigates the behaviour of robust residuals. The emphasis here is on probability plots. Section 2 presents some characteristics and problems associated with least-squares residuals. Section 3 presents a robust estimation procedure which yields residuals with properties that are easily studied. The shape of probability plots of residuals from this estimation procedure is discussed and illustrated. In Section 4, plots of residuals against other variables are considered.

2. PROBABILITY PLOTS OF LEAST SQUARES RESIDUALS

Probability plots are commonly generated for residuals from linear models. See for example Daniel and Wood (1972). These displays are produced by plotting ordered residuals $r_{(i)}$ vs values E_i, representative of a reference distribution F. These representative values might be expected values of order statistics from the distribution F. This suggestion, while common, has little to recommend it over others. The order statistic distributions are not symmetric and so medians would seem to be more easily interpreted. Computational considerations often lead to the use of

$$E_i = F^{-1}((i - \delta)/(n + 1 - 2\delta)) \quad ;$$

$\delta = 0, \frac{1}{2}, \frac{1}{3}$ being common choices. The Gaussian distribution is normally used for F.

If the residuals r_i are a sample of size n from the distribution F, such a plot tends to a straight line configuration. Departures from this configuration are taken as evidence of 'outliers', model inadequacies or incorrect distributional assumptions. Detection of these departures usually results in modifications to the comtemplated model.

However residuals are not a simple random sample from the distribution F. The fitting of parameters tends to perturb the typical linear reference configuration. A more serious problem occurs when the distribution of disturbances is non-Gaussian. In this case the fitting of parameters tends to wash out the evidence of non-Gaussianity contained in the residuals.

Consider the model

$$\underset{\sim}{y} = X\underset{\sim}{\beta} + \underset{\sim}{e}$$

where $\underset{\sim}{y}$ is an $n \times 1$ vector of variables, X is a matrix of independent variables of rank p, $\underset{\sim}{\beta}$ is a vector of unknown parameters, and $\underset{\sim}{e}$ is a vector of disturbances. Let W be the matrix which produces the least-squares residuals $\underset{\sim}{r}$

$$\underset{\sim}{r} = W\underset{\sim}{y} \ .$$

If the disturbances $\underset{\sim}{e}$ may be described as a random sample with a distribution F(e), moments and other distributional properties of the residuals $\underset{\sim}{r}$ may be calculated.

Let $\kappa_{im}(e)$ and $\mu_{im}(e)$ be the mth cumulant and moment of e_i respectively. (The subscript i is redundant here but will be useful later.) The cumulants of r_i are defined analogously and satisfy

$$\kappa_{im}(r) = \sum_j w_{ij}^{\ m} \kappa_{im}(e)$$

From this, the moments of r_i may be found using the usual relations between cumulants and moments:

$$\mu_{im}(r) = \mu_{im}(\underset{\sim}{\kappa}_{i.}(r)) \ .$$

We may consider the r_i as having a distribution G . Although the residuals are not independent this does not affect the averages and expectations considered below.

The mth sample moment has expectation $\mu_m(\underset{\sim}{r})$ given by

$$\mu_m(\underset{\sim}{r}) = n^{-1} \sum_i \mu_{im}(r)$$

These moments may be used to calculate the corresponding cumulants

$$\kappa_m(\underset{\sim}{r}) = \kappa_m(\mu_.(\underset{\sim}{r}))$$

Since, for the purposes considered here, cumulants describe the shape of the distribution, the matrix W defines a mapping from the shape of F(e) to the shape of G(r):

$$F \overset{W}{\rightarrow} G$$

The following examples illustrate the effect of fitting parameters in a 2^5 experiment, main effects and first order interactions being fitted. Data were generated from a χ^2_4 distribution. In this discussion all distributions are adjusted to have mean 0, variance 1. Figure 1 is a probability plot of the disturbances. The theoretical cumulants for these disturbances are:

Theoretical: 0, 1, 1.4, 3, 8.5, 30,...

Sample : 0, 1, 0.9, -0.1, -4.3, -10.4,...

Figure 2 is a probability plot of the least squares residuals from the fitted model. The standardized cumulants for these residuals are:

Theoretical: 0, 1, 0.06, 0.03, 0.008, .004,...

Sample : 0, 1, 0.2, -0.6, -0.8, 1.0,...

The fitting of parameters has reduced markedly the apparent departure from the Gaussian straight line configuration. The least-squares residuals can give little indication when corrective action is required.

A number of changes may be made in the plotting procedure to allow for this effect.

The relation

$$\kappa(e) \overset{W}{\rightarrow} \kappa(r)$$

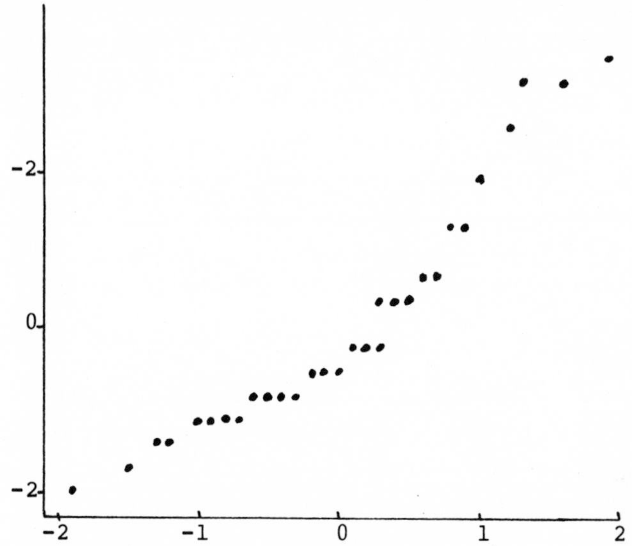

Figure 1

Probability plots of disturbances - standardized χ_4^2 variables.

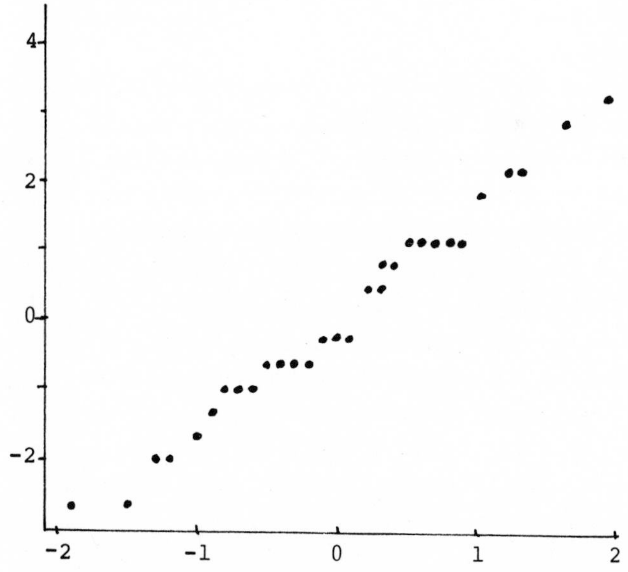

Figure 2

Probability plots of residuals from data of Figure 1.

may be inverted and the sample cumulants of r used to estimate the cumulants of e. This is analogous to the development in Anscombe (1961). These estimated cumulants $\hat{\kappa}(e)$ may be used to estimate the shape of the distribution $\hat{F}(e)$. In particular a Cornish-Fisher expansion, may be used to transform $F^{-1}(p)$, into the percentage points of the estimated distribution $\hat{F}^{-1}(p)$. These can be plotted against the values of $F^{-1}(p)$. This procedure requires up to six moments and can be safely used only on very large problems with $\hat{F}$ close to F where the Cornish-Fisher expansion is reasonable.

An alternative procedure, is to consider monotonic transformation of $r \to r^*$ to match say the first three moments of $\hat{F}$ and then produce a probability plot of r^*. This approach is developed in Andrews and Pregibon (1978).

The discussion thus far has been about least-squares residuals. It is important to understand the corresponding effects for residuals from robust fits. To facilitate the understanding of these residuals a particular robust estimation procedure is introduced in the next section.

3. TRIMMED BETAS

This section describes a simple analogue of trimmed means, useful for studying residuals from a robust estimation procedure. It is not recommended for application. Its only useful attributes are that it is in some sense robust and that it leads to residuals with convenient properties. The estimator is defined by the following algorithm based on iterated weighted least-squares.

1) Initialize weights
2) Calculate the residuals using weighted least-squares
3) Define weights to be 0 or 1
4) Iterate steps 2 and 3 until a fixed number $[\alpha n]$ of the weights are 0 for both the corresponding largest and smallest $[\alpha n]$ residuals, the remaining weights being 1.

Assume for the moment that a precise specification of this algorithm exists leading to convergence for the X matrix involved. The α-trimmed mean is the simplest example of this estimate in which case steps 2,3 and 4 are trivial and require <u>no</u> iteration.

Let $r(\alpha)$ denote the residuals from this procedure, $r(0)$ being the least-squares residuals. Let $D = D(\alpha)$ be the diagonal matrix of weights. Thus the estimate, $\hat{\beta}(\alpha)$, satisfies

$$(X'DX)\hat{\beta}(\alpha) = X'D\,y$$

and the residuals are defined by a corresponding W matrix

$$r(\alpha) = (W(\alpha))y$$

where

$$W(\alpha) = [I - X(X'DX)^{-1}X'D]$$

The moments of $r(\alpha)$ may be estimated using the methods of the previous section.

The sample moments of e, r, and $r(\alpha)$ and hence the shape of the corresponding empirical distribution functions may be studied by considering each of these as linear combinations:

$$e_i = \sum_j v_{ij}\, e_{ij} \quad ; \quad v_{ii} = 1 \quad ; \quad v_{ij=i} = 0$$

$$r_i = \sum_j w_{ij}\, e_j$$

$$r(\alpha)_i = \sum_j w(\alpha)_{ij}\, e_j$$

Suppose $F(e)$ represents a distribution with positive third moment. This will typically result in a positive sample third moment. A large part of this third sample moment comes from the largest of the e_i.
Figure 3 exhibits the weights v_{ij}, w_{ij}, $w(\alpha)_{ij}$ plotted vs e_i for the largest $e_{(i)}$.

The typically negative correlations (w_{ij}) between least squares residuals tend to wash out the contribution of e_i to the third moment $\Sigma r_i{}^3$. The 0 correlations among the $r(\alpha)_j$ corresponding to the largest and smallest e_i preserve the contribution of these e_i to the third moment. Thus the moments of $r(\alpha)$ will resemble more closely those of e than will the moments of r.

This observation has an asymptotic analogue.

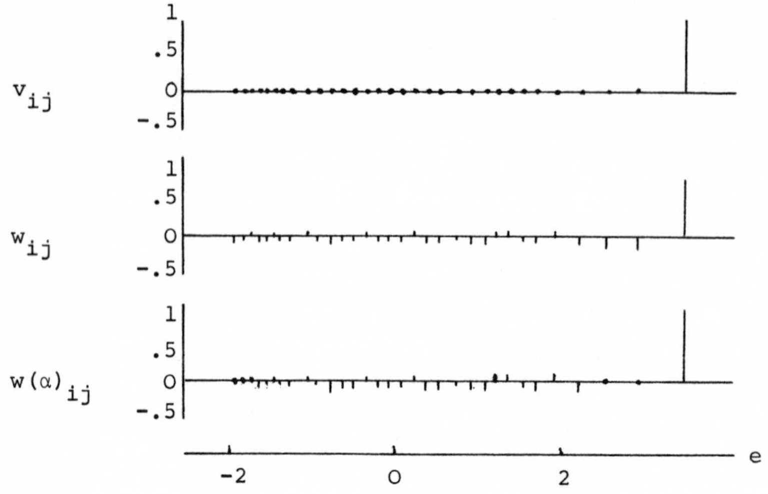

Figure 3

A comparison of coefficients of $\mathbf{e}_{(i)}$ for calculating moments of $r_{(n)}$.

v_{ij} - fitting by parameters

w_{ij} - fitting by least squares

$w(\alpha)_{ij}$ - fitting by trimmed Betas

Consider a density function f with support partitioned as in Figure 4

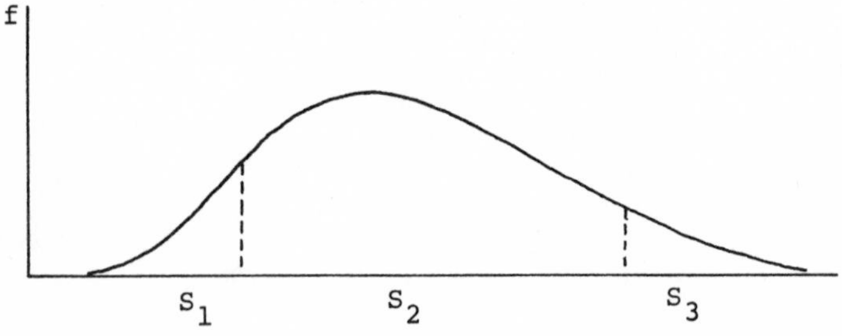

Figure 4

A density function with support S_1, S_2, S_3

such that $P(S_1) = P(S_3) = \alpha$ and $P(S_2) = 1 - 2\alpha$.

Let f_k $k = 1,2,3$ be the corresponding conditional densities

$$f_k(e) = f(e)\ I(S_k)/P(S_k)$$

where I is the indicator function.
Let μ_m^k be the moments of f_k. Then the moments of $F(e)$ satisfy

$$\mu_m(e) = \sum_k P(S_k)\ \mu_m^k(e)\ .$$

A Random sample $e_1,\ldots,e_n$ from f may be generated by first generating a multinomial variable n_1,n_2,n_3 corresponding to probabilities α, $1 - 2,\alpha$ and then generating n_k samples from f_k. Assume that the trimmed β estimate is consistent so that

$$r(\alpha)_i = e_i + O(n^{-\frac{1}{2}})\ .$$

The residuals are associated with weights according to the size of the residual. Thus the residuals are associated through the estimation procedure, with the sets S_k. Let $N_{k\ell}$ be the number of residuals corresponding to disturbances e coming from support S_k which have been assigned a weight corresponding to support S_ℓ. Clearly the fraction of misclassifications goes asymptotically to 0:

$$\sum_{k\neq\ell} N_{k\ell} = O(\sqrt{n})\ .$$

The mapping $\kappa(e) \overset{W}{\rightarrow} \kappa(r)$ of Section 2 was expressed in terms of the cumulants of e. These cumulants are known subject to the errors of misclassification which are asymptotically small. Thus, conditionally, given $W(\alpha)$, the sample cumulants of $r(\alpha)$ may be calculated. The mapping

$$\kappa(e) \overset{W(\alpha)}{\rightarrow} \kappa(r)$$

may be inverted to yield estimated cumulants of e and hence the estimated shape of the probability plot of e.

The shape of residual plots may be studied using the following functions

$\underset{\sim}{\mu} \to \underset{\sim}{\kappa}$	calculate cumulants from moments
$\underset{\sim}{\kappa} \to \underset{\sim}{\mu}$	calculate moments from cumulants
$\underset{\sim}{r} \to \underset{\sim}{\kappa}(\underset{\sim}{r})$	calculate sample cumulants of $\underset{\sim}{r}$
$f(e) \to \kappa(e)$	calculate cumulants of $f(e)$
$z,\kappa(e) \to e = F^{-1}(\Phi(z))$	Cornish-Fisher expansion
$e,\kappa(e) \to z = \Phi^{-1}(F(e))$	Fisher-Cornish Expansion
$y,X \to \hat{\beta}(\alpha),W(\alpha),r(\alpha)$	trimmed β estimate
$\underset{\sim}{\kappa}(\underset{\sim}{e}),W(\alpha) \to \kappa(r(\alpha))$	calculate residual cumulants
$\kappa(r),W(\alpha) \to \kappa(e)$	estimate cumulants of e from sample cumulants of r.

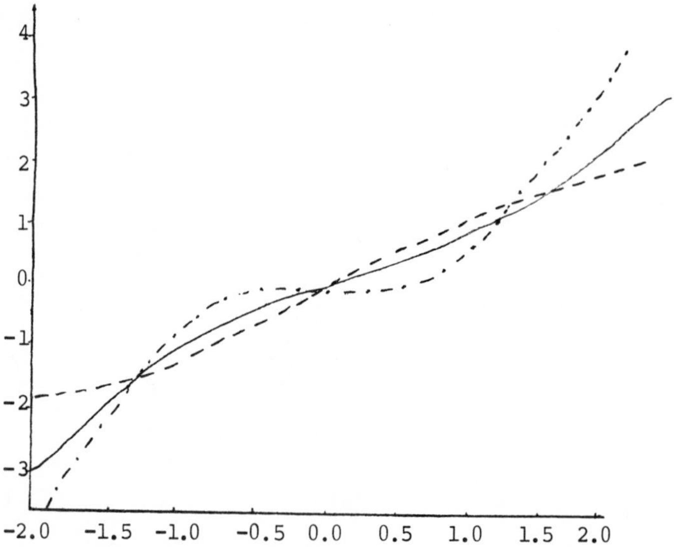

Figure 5

A comparison of the shape of probability plots of least squares residuals ------ and robust residuals -·-·-·- with the plot based on the slightly long-tailed disturbances ————.

Figure 5 exhibits the shape of the probability plots for e, r, and $r(\alpha)$ for a 2^5 experiment with main effects and first order interactions fitted. The distribution of e was non-Gaussian, given by

$$F^{-1}(p) = p^{\lambda} - (1 - p)^{\lambda} \quad \lambda = -0.1$$

The plot of the least squares residuals is very straight. The evidence of non Gaussianity has been lost. The trimmed β residuals $\alpha = 3/32$ resemble much more closely the shape of the plot of e. Some over-correction has occurred. This is characteristic of many robust procedures. Over-sensitivity is often the price of robustness.

4. PLOTS AGAINST OTHER VARIABLES

Robust residuals tend to point more clearly to departures from distributional assumptions. There is some interest also in the effects of model misspecification. It is clear that common residual displays are quite insensitive to some non linear inadequacies of the model.

Consider the model as before

$$\underset{\sim}{y} = X\underset{\sim}{\beta} + \underset{\sim}{e} \tag{4.1}$$

and data generated by

$$\underset{\sim}{y} = X\underset{\sim}{\beta} + f(\underset{\sim}{x}^*) + N(0,1) \tag{4.2}$$

The least squares residuals given by $\underset{\sim}{r} = W\underset{\sim}{y}$ satisfy

$$\underset{\sim}{r} = Wf(\underset{\sim}{x}^*) + W\underset{\sim}{e}$$

That x^* is not yet included in the model may be due in part to its non-linear contribution. Much common practice suggests that a plot of r vs x^* will exhibit this dependence. However there is no reason why the residuals should look anything like $f(x^*)$. Indeed since both X and $f(x^*)$ are related to y, $Wf(x^*)$ is likely to be very different from $f(x^*)$. A fuller discussion of this with examples of some clarifying procedures is discussed in Andrews and Pregibon (1978).

What is the effect of using a robust estimation procedure on these plots? Since

$$r(\alpha) = W(\alpha)y = (I - X(X'DX)^{-1}X'D)y$$

the effect of changing from least-squares to a trimmed β estimate can be seen from changes in the elements of W. The largest residuals will tend to be associated with the largest values of $Wf(x^*)$. The covariances among the residuals with weight 0 is 0. The covariances with other residuals are typically increased in magnitude. The net effect is to typically increase the absolute value of the residuals with 0 weight and decrease that of the others. Thus the effect of increasing α is to increase the magnitude of large residuals. Since the magnitude depends on the size of $Wf(x^*)$ and not on x^* this stretching will tend to make departures more visible. However if $Wf(x^*)$ is very different from x^* this departure may appear as noise and will not be detected.

Consider the model

$$y = 1\beta_o + x_1\beta_1 + \text{noise} \tag{4.3}$$

and data generated by

$$y = x_1 + 0.05x_2^2 + N(0,1) \tag{4.4}$$

where $x_1 = 1,1,2,2,\ldots 10,10$ and $x_2 = 1,2,\ldots,20$

Figure 6 is a plot of the least-squares residuals from (4.3) vs x_2. The quadratic dependence is obscured here. Figure 7 is a plot of $r(.1)$ the residuals from a 10% trimmed β estimate. Again the quadratic dependence is far from clear. Robust estimation has not helped to clarify the inadequacy of the model.

5. CONCLUSION

Least-squares residuals, when plotted simply, do not clearly reveal the nature of the inadequacies in the contemplated model. This effect is strong when many parameters are fitted. Probability plots of robust residuals tend to

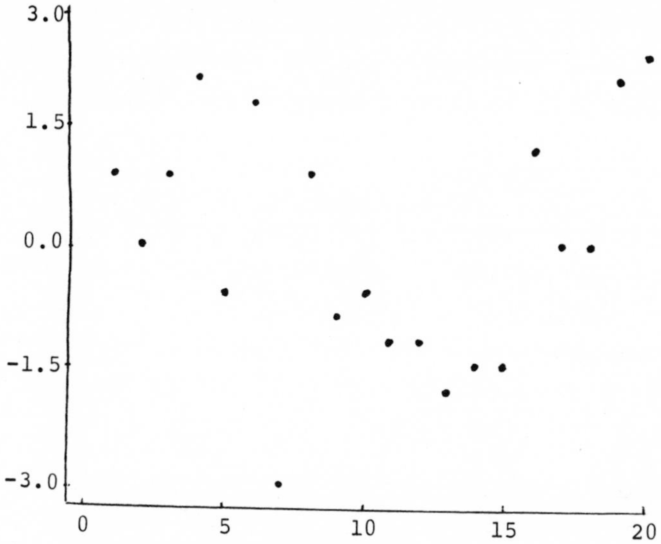

Figure 6

Least-squares residuals plotted against the variable x_2. These residuals do not show the dependence $0.05x_2^2$ of y on x_2.

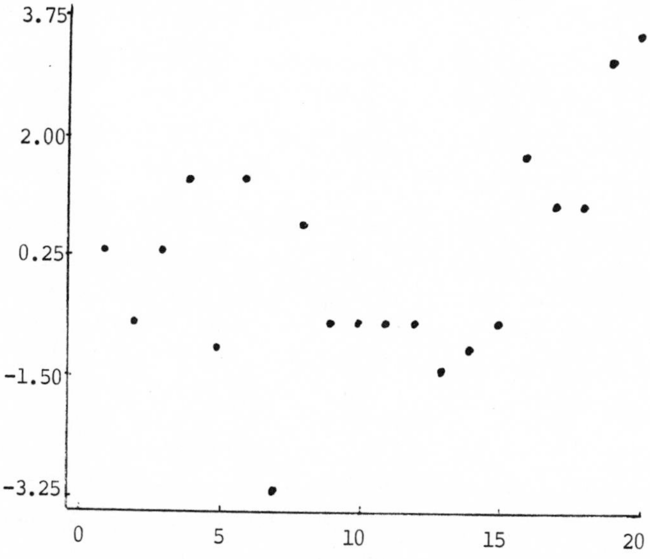

Figure 7

Robust residuals plotted against the variable x_2 as in Figure 6. Robust estimation has not clarified this plot.

show these departures more clearly. The effect of fitting parameters on either set of residuals may be studied using the cumulants of the disturbances. The cumulants of the distribution of residuals may be caluculated and used in a Cornish-Fisher expansion to estimate the shape of the distribution of residuals. The tendency of least-squares residuals toward a Gaussian distribution is not exhibited by robust residuals.

Plots of least-squares residuals against other variables do not clearly exhibit non-linear dependencies on variables omitted from the model. Robust residuals do not improve these plots.

REFERENCE

1. Andrews, D.F. and Pregibon, D. (1978). Claryifying displays of residuals. University of Toronto, Department of Statistics, Technical Report.

2. Anscombe, F.J. (1961). Examination of residuals, Proceedings of the Fourth Berkeley Symposium. J. Neyman ed., University of California Press.

3. Daniel, C. and Wood, F.S. (1971). Fitting Equations to Data, Wiley, New York.

4. Mallows, C.L. (1969). Techniques for non-standard order statistics, Bull. I.S.I. 43, 1969, Book 2, 154-156.

5. ____________. (1977). The role of robustness in the analysis of large data-sets, presented IMS Conference on Large Data-Sets, Dallas, February 23-5, 1977.

This research was supported in part by The Canada Council while the author was visiting Imperial College. I am grateful to Daryl Pregibon who contributed much to the development of this work and to Susan Pregibon who prepared the article for publication.

Department of Statistics
University of Toronto
Toronto, Ontario
Canada M5S 1A1

Robust Smoothing

Peter J. Huber

1. INTRODUCTION

Suppose that a curve is sampled without repetition at discrete points, and that the measurements contain observational errors. Then a smoothing procedure is needed in order to:

- check the data for outliers,
- substitute missing values,
- interpolate between grid points,
- produce nicer graphs by reducing random fluctuations.

For several of the above purposes it is desirable that the smoothing is robust, i.e. not overly affected by occasional outliers in the data. This can be achieved in an informal or semiformal way by first smoothing (in some arbitrary fashion), then throwing out data points deviating too much from the smoothed curve, and finally smoothing again. However, both the performance and the side effects of such an approach are difficult to appraise, and during a recent study of children's growth data (Largo et al. (1978)) it occurred to us that a more rigidly formalized approach would have definite advantages.

ISBN 0-12-438150-2

First, a number of desirable properties for a good smoother shall be listed. Let X_t be the time series to be smoothed, let S be the smoothing operator and let $Z_t = (SX)_t$ be the smoothed series. The domain of t need not be the same for Z_t as for X_t.

Then we should like S to satisfy the following.

(i) It must be tuneable from no smoothing ($Z_t = X_t$ at the sample points) to extreme smoothing (e.g. Z_t linear).

(ii) It must contain a robustness parameter and be tuneable from no robustness (ordinary linear smoothing) to high robustness. Robustness must at least include resistance toward isolated outliers.

(iii) Genuine discontinuities, in particular sudden, permanent changes in the mean value of X_t, should not be smeared over a wide range. This might be achieved through a range parameter r, so that Z_t depends only on the X_u with $|t-u| \leq r$.

(iv) Very smooth functions should be reproduced exactly (e.g. SX = X if X is a low order polynomial in t).

(v) Any robust smoother is non-linear by necessity, and the spectrum of SX may show some weird effects of non-linearity. The smoother should be simple enough so that these effects can be anticipated and kept under control.

(vi) The smoothed curve should be reasonably simple to calculate.

Smoothers based on moving medians (Tukey 1977) are specifically geared towards requirement (iii), but they are not easily tuneable. In particular, it is not possible to tune their smoothness and robustness indepently of each other. Moreover, it is quite difficult to relate them to the linear smoothers whose behavior we believe to understand (Mallows 1977).

Our viewpoint on smoothing is different from that of classical filtering theory (whose principal goal is to pick out a known signal hidden behind known noise): smoothers are crude low-pass filters to be used as data analytic tools before much is known about the stochastic structure of X. They should therefore be defined with as little recourse to this unknown structure as feasible, but on the other hand, their behavior should be known for a wide variety of processes.

The theoretically cleanest approach to linear smoothing is through splines (Reinsch 1967): minimize the mean square of the second (or of a higher order) derivative of Z:

$$\text{ave}\ \{(Z_t'')^2\} \tag{1.1}$$

subject to a side condition of the form

$$\text{ave}\ \{(X_t - Z_t)^2\} \leq \text{const.} \tag{1.2}$$

The averages are taken over a suitable range of t-values (which usually is continuous for (1.1), but discrete for (1.2)). It is often convenient to combine (1.1) and (1.2) into a single expression with the aid of a Lagrange multiplier.

This approach can be robustified very easily: we simply replace the square in (1.2) by a less rapidly increasing function ρ. Past experience with location and regression estimates suggests that ρ should be chosen convex with a bounded derivative $\psi = \rho'$, for example

$$\begin{aligned} \rho(x) &= \frac{1}{2}x^2 && \text{for} \quad |x| \leq c \\ &= c\,|x| - \frac{1}{2}c^2 && \text{for} \ |x| > c. \end{aligned} \tag{1.3}$$

The constant c regulates the degree of robustness; good choices for c are between 1 or 2 times the standard deviation of the individual observations X_t. We shall comment later on other, in particular non-convex choices for ρ.

Then the smoothed process Z can be defined by the property that it minimizes

$$\text{ave}\ \{\rho(X_t - Z_t)\} + \frac{1}{2}b^4\ \text{ave}\ \{(Z_t'')^2\}, \tag{1.4}$$

where $\frac{1}{2}b^4$ is a Lagrange multiplier, regulating the degree of smoothness.

I am sure that the above simple robustification (1.4) of smoothing splines must have occurred to several people independently; but at present, I am only aware of two such studies: Lenth (1977) and Gasser and Krumm (1978).

2. CIRCULAR SMOOTHING

We first discuss the simplest case: assume that the domain of t is the set of integers modulo N, both for X and for Z. Of course, the second derivative Z'' then must be replaced by a second difference quotient

$$\Delta^2 Z_t = Z_{t+1} - 2Z_t + Z_{t-1}, \tag{2.1}$$

and we have to minimize

$$\text{ave}\,\{\rho(X_t - Z_t) + \frac{1}{2}\, b^4\, (\Delta^2 Z_t)^2\}\ . \tag{2.2}$$

This is a convex function of $Z = (Z_1,\ldots,Z_N)$, and a necessary and sufficient condition for a minimum is that (2.2) remains stationary under infinitesimal variations of Z:

$$\text{ave}\,\{-\psi(X_t - Z_t)\,\delta Z_t + b^4\,\Delta^2 Z_t \cdot \Delta^2\,\delta Z_t\} = 0. \tag{2.3}$$

"Integration" by parts yields

$$\text{ave}\,\{-\psi(X_t - Z_t)\,\delta Z_t + b^4\,\Delta^4 Z_t \cdot \delta Z_t\} = 0. \tag{2.4}$$

This must hold for all choices of δZ, hence we obtain as a necessary and sufficient condition for the minimum

$$b^4\,\Delta^4 Z_t = \psi(X_t - Z_t). \tag{2.5}$$

We first consider the special case $\psi(x) = x$, corresponding to linear smoothing. Then (2.5) becomes

$$Z_t + b^4\,\Delta^4 Z_t = X_t, \tag{2.6}$$

or, written in operator form

$$(1 + b^4\,\Delta^4)Z = X, \tag{2.7}$$

with the formal solution

$$Z = (1 + b^4\,\Delta^4)^{-1}X \tag{2.8}$$

An explicit solution can be found easily by Fourier methods. Put

$$\hat{X}_q = \frac{1}{N} \sum_t X_t \exp(-2\pi iqt/N) \tag{2.9}$$

then

$$X_t = \sum_q \hat{X}_q \exp(2\pi iqt/N) \tag{2.10}$$

and similarly $\hat{Z}_q$; all sums are over the full period $0,\ldots,N-1$.

Put

$$\Delta Z_t = Z_{t+\frac{1}{2}} - Z_{t-\frac{1}{2}} \tag{2.11}$$

(for half-integral values of t), then the Fourier transform of ΔZ is

$$(\Delta Z)\hat{}_q = 2\sin\left(\frac{\pi q}{N}\right) \hat{Z}_q, \tag{2.12}$$

and hence

$$(\Delta^4 Z)\hat{}_q = \left(2\sin \frac{\pi q}{N}\right)^4 \hat{Z}_q \ . \tag{2.13}$$

Thus we obtain from (2.6) or (2.7) that

$$\hat{Z}_q = \frac{1}{1 + \left(2b \sin \frac{\pi q}{N}\right)^4} \hat{X}_q \ , \tag{2.14}$$

and that

$$Z_t = \sum_u a_{N,u} X_{t-u} \ , \tag{2.15}$$

with

$$a_{N,t} = \frac{1}{N} \sum_q \frac{\cos\left(2\pi \frac{qt}{N}\right)}{1+\left(2\,b\sin \frac{\pi q}{N}\right)^4} \ . \tag{2.16}$$

For $N \to \infty$ these coefficients converge quite rapidly to

$$a_t = \frac{1}{\pi} \int_0^{\pi} \frac{\cos(\omega t)}{1 + \left(2\,b\sin \frac{\omega}{2}\right)^4} d\omega \ . \tag{2.17}$$

Table 1 lists some numerical results; they seem to indicate that b-values below 1 ordinarily will give too little, values above 4 too much smoothness.

Table 1

	b = 1	b = 2	b = 4
a_o	.3882	.1820	.0891
a_1	.2366	.1606	.0863
a_2	.0879	.1221	.0799
a_3	.0113	.0818	.0712
a_4	-.0118	.0478	.0614
a_5	-.0115	.0228	.0512
a_6	-.0059	.0066	.0415
a_7	-.0018	-.0025	.0324
a_8	.0000	-.0065	.0244
a_9	.0004	-.0074	.0175
a_{10}	.0003	-.0065	.0117
a_{11}	.0001	-.0049	.0071
a_{12}		-.0033	.0034
a_{13}		-.0019	.0007
a_{14}		-.0009	-.0013
a_{15}		-.0003	-.0026
a_{16}		.0001	-.0033
a_{17}		.0003	-.0037
a_{18}		.0003	-.0038
a_{19}		.0003	-.0036
a_{20}		.0002	-.0033
a_{21}		.0001	-.0030
a_{22}		.0001	-.0025
a_{23}		.0000	-.0021
a_{24}			-.0017
a_{25}			-.0013
a_{26}			-.0010
a_{27}			-.0007
a_{28}			-.0005
a_{29}			-.0003
a_{30}			-.0001

We note in passing that

$$\sum a_t = 1$$
$$\sum t^k a_t = 0 \qquad \text{for } k = 1,2,3. \tag{2.18}$$

In other words, smoothing with coefficients a_t will reproduce any polynomial X_t of degree ≤ 3.

The coefficients a_t have an infinite range; if we prefer a smoother with finite range, we can in principle solve the minimum problem (2.2) by first substituting (2.15) for Z_t, and then minimizing numerically, subject to the side conditions (2.18) and $a_t = 0$ for $|t|>r$. But in practice, it would seem to be adequate to truncate the sequence a_t at a suitable t.

We now return to the general robust smoothing problem (2.5) and write it as

$$(1 + b^4 \Delta^4)Z = Z + \psi(X-Z). \tag{2.19}$$

This problem is closely related to robust regression and can be solved by methods borrowed from there (the H-algorithm of Huber and Dutter (1974) and Dutter (1977)). We choose some starting curve $Z^{(o)}$ (see below) and define recursively.

$$Z^{(m+1)} = (1 + b^4\Delta^4)^{-1} (Z^{(m)} + \psi(X-Z^{(m)})), \tag{2.20}$$

i.e., if we suppress the index N from now on:

$$Z^{(m+1)}_t = \sum_u a_u \cdot (Z^{(m)}_{t-u} + \psi(X_{t-u} - Z^{(m)}_{t-u})). \tag{2.21}$$

If we choose $Z^{(o)} = X$ as the starting curve then $Z^{(1)}$ is the ordinary, linearly smoothed version. For some purposes, $Z^{(o)} = 0$ is a convenient choice. Practical considerations, in particular experience with one-step estimates of location (Andrews et al. (1972)), might suggest to take for $Z^{(o)}$ the result of a moving median (of 3 to 5 elements) applied to X, so that already the starting curve is insensitive to outliers.

The proof that $Z^{(m)}$ converges to a solution of (2.5) is also borrowed from robust regression. Assume $0 \leq \rho'' \leq 1$. We first show that the step (2.21) decreases (2.2). Replace the term $\rho(X_t - Z_t)$ in (2.2) by a quadratic function of Z_t, namely

$$U_t(Z_t) = \rho(X_t - Z_t^{(m)}) - \psi(X_t - Z_t^{(m)})(Z_t - Z_t^{(m)}) + \frac{1}{2}(Z_t - Z_t^{(m)})^2 \quad (2.22)$$

We have $U_t(Z_t) \geq \rho(X_t - Z_t)$ for all Z_t, with equality for $Z_t = Z_t^{(m)}$, and one easily checks that $Z^{(m+1)}$ minimizes the modified (2.2). Hence the step (2.21) strictly decreases the original (2.2), unless $Z^{(m+1)} = Z^{(m)}$ (and then $Z^{(m)}$ is an exact solution of (2.5)). By a compactness argument one shows that the sequence $Z^{(m)}$ has at least one accumulation point, and that every accumulation point is a solution of (2.5). If the latter is unique, the proof is terminated.

In the rather pathological case where the solution of (2.5) is not unique, it is still possible to show that $Z^{(m)}$ always converges to some solution.

3. MORE GENERAL CASES

It should be evident from the preceding section how more general robust smoothing problems (1.4) can be treated. To be specific, assume that X has been observed at the not necessarily equidistant points $t_1, \ldots, t_N$, and that Z should be determined on the interval t_1, t_N. Then in the classical, linear case ($\rho(x) = \frac{1}{2} x^2$), the solution is a cubic smoothing spline with knots at the t_i. The values Z_t depend linearly on $X_{t_1}, \ldots, X_{t_N}$, say

$$Z_t = \sum_i K_i(t) X_{t_i} . \quad (3.1)$$

The robustified version can be solved by essentially the same iterative procedure that was suggested in the preceding section: assume that $Z^{(m)}$ is the current trial curve and replace X_{t_i} by

$$X_{t_i}^{(m)} = Z_{t_i}^{(m)} + \psi(X_{t_i} - Z_{t_i}^{(m)}). \tag{3.2}$$

For example, if ψ is determined by (1.3), we have $X_{t_i}^{(m)} = X_{t_i}$ if $|X_{t_i} - Z_{t_i}^{(m)}| \leq c$, and otherwise $X_{t_i}^{(m)} = Z_{t_i}^{(m)} \pm c$. Then apply the smoothing process to $X^{(m)}$ and iterate:

$$Z_t^{(m+1)} = \sum_i K_i(t)\, X_{t_i}^{(m)} \tag{3.3}$$

4. THE CHOICE OF ψ.

Assume that the X_t are the sum of a very smooth function plus independent identically distributed random variables with probability density f, and that we are concerned with the asymptotic situation where there is so much smoothing that the statistical variability of Z_t is negligible against that of X_t. (Admittedly, this case is not very realistic - one rarely will want to smooth that much.) Then the problem reduces essentially to that of estimating a location parameter, and just as there the asymptotically optimal (efficient) choice for ψ is

$$\psi = -(\log f)' \tag{4.1}$$

Keeping other things equal, this will minimize the asymptotic variance of Z_t.

If f is only approximately known, we can determine an asymptotic minimax solution by taking for f a least informative density. For ε-contaminated normal distributions we end up with $\rho = -\log f$ of the form (1.3), just as in the location case. The corresponding smoothers treat outliers by metrical Winsorization, cf. (3.2) and the following remarks. It is possible, at a very slight increase of the minimax risk, to improve the performance for very long tails, if one uses a "redescending" ψ-function (cf. Andrews et al. (1972)). However, great care is needed: the iterative procedures may easily get stuck in a local minimum of (1.4) if ρ is not convex.

5. ESTIMATING A SCALE PARAMETER

The robust smoothers discussed so far are not scale invariant. In practice, we should therefore estimate a scale parameter S from the sample and replace $\psi(X_t - Z_t)$ by

$$\psi\left(\frac{X_t - Z_t}{S}\right) S \tag{5.1}$$

in (2.5) and all other formulas.

For instance, we may use the median absolute deviation

$$S = \operatorname{med}\{|X_t - Z_t|\} \tag{5.2}$$

Unfortunately, the convergence of the procedure which iterates simultaneously on Z and on S (by inserting $Z^{(m)}$ in (5.2)) has not yet been proved.

Or, we may estimate S by solving

$$\operatorname{ave}\left\{\psi\left(\frac{X_t - Z_t}{S}\right)^2\right\} = \beta \tag{5.3}$$

for S, with the $\psi = \rho'$ of (1.3), for some constant β (see below). Here it can be shown that alternating between steps (2.20) and

$$(S^{(m+1)})^2 = \frac{1}{\beta} \operatorname{ave}\left\{\left[\psi\left(\frac{X_t - Z_t^{(m)}}{S^{(m)}}\right) S^{(m)}\right]^2\right\} \tag{5.4}$$

gives convergence.

A reasonable choice for β would seem to be

$$\beta = E_\Phi(\psi^2) = \frac{1}{\sqrt{2\pi}} \int \psi(x)^2 \, e^{-x^2/2} \, dx. \tag{5.5}$$

Then, if the X_t are i.i.d. normal $\mathcal{N}(\mu, \sigma^2)$, and if $b \to \infty$, S is a consistent estimate of σ. It remains to be investigated whether the same choice of β is also appropriate for low values of b (little smoothing).

6. THE CHOICE OF THE SMOOTHNESS PARAMETER b.

It may be desirable to determine the value of the smoothness parameter from the data. The following approach, based on cross-validation techniques (Wahba and Wold (1975)) has been fairly successful in our experience with non-robust smoothing splines (Stützle (1977), Largo et al. (1978)). Let Y_t be the interpolated value at t, i.e. the smoothed value calculated on the basis of the observations X_u, $u \neq t$. Do this for all data points t and adjust b such that the average squared prediction error ave $\{(X_t - Y_t)^2\}$ is made smallest.

In detail, this works as follows. Write (2.15) as

$$Z_t = a_o X_t + \sum_{u \neq 0} a_u X_{t-u} \tag{6.1}$$

We define the interpolated value Y_t to be that value at the point t which is reproduced when (6.1) is applied to it. That is, Y_t can be found from the equation

$$Y_t = a_o Y_t + \sum_{u \neq 0} a_u X_{t-u}. \tag{6.2}$$

Thus

$$Y_t = (\sum_{u \neq 0} a_u X_{t-u})/(1 - a_o). \tag{6.3}$$

(Note that $\sum_u a_u = 1$).

Then the residuals relative to the smoothed and to the interpolated values satisfy

$$X_t - Z_t = (1-a_o)X_t - \sum_{u \neq 0} a_u X_{t-u} = (1-a_o)(X_t - Y_t). \tag{6.4}$$

We find that we have to minimize

$$\text{ave } \{(X_t - Y_t)^2\} = \frac{1}{(1-a_o)^2} \text{ ave } \{(X_t - Z_t)^2\} . \tag{6.5}$$

Clearly, because of the averaging over t, any such procedure will oversmooth when the true underlying curve is very smooth over most of its range, apart from a rough spot of physical interest ("a straight line with a bump"). With robust smoothing, the problem will become even worse.

Also another potential trouble should be pointed out. We note that our spline smoothers act as low pass filters; for $b \gtrsim 2$ they have a fairly sharp cut-off frequency ω_o determined by

$$2b \sin \frac{\omega_o}{2} = 1 \tag{6.6}$$

(i.e. $\hat{Z}_q \approx \hat{X}_q$ or $\hat{Z}_q \approx 0$ according as $2\pi q/N < \omega_o$ or $> \omega_o$).

If we introduce the spectral density

$$f(\omega) = \frac{N}{2\pi} E(|\hat{X}_q|^2), \qquad \omega = \frac{2\pi q}{N}, \tag{6.7}$$

we obtain the following approximate representation:

$$\text{ave}\,\{(X_t - Y_t)^2\} \approx \frac{2\int_{\omega_o}^{\pi} f(\omega)\,d\omega}{\left(1 - \frac{\omega_o}{\pi}\right)^2} \tag{6.8}$$

This formula tells us at a glance that the smoothing parameter will be very poorly determined if the spectral density is approximately

$$f(\omega) \sim \left(1 - \frac{|\omega|}{\pi}\right)^+ . \tag{6.9}$$

Different samples from the same process will then lead to widely different b-values, with the disadvantage that the smoothed versions are not comparable among themselves.

It appears that no automatic, data-driven method for selecting b will ever be foolproof, and a visual inspection, which can take into account our informal, intuitive knowledge about the underlying structure of X, will always remain advisable.

With these precautions in mind, I would tentatively propose to robustify the above by minimizing, instead of (6.5),

$$\frac{1}{(1-a_o)^2} \operatorname{ave}\left\{\left[\psi\left(\frac{X_t - Z_t}{S}\right) S\right]^2\right\}, \tag{6.10}$$

where the scale parameter S has been determined by one of the methods suggested in the preceding section.

7. LEAKAGE

We have already mentioned that non-linear smoothers may create some awkward artifacts in the frequency domain: strong high frequency components in the spectrum of X (which themselves may be smoothed away!) can combine and "leak" a noticeable low frequency contribution into the spectrum of Z. Not all of these effects are necessarily undesirable; for instance if we smooth a process of the form depicted in Fig. 1 with ave $\{X_t\} = 0$,

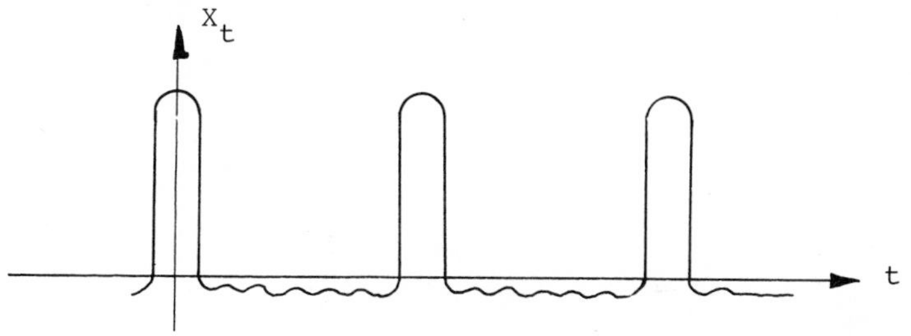

Fig. 1

then robust smoothing should crop the narrow positive peaks and return a smoothed version with ave $\{Z_t\} < 0$. In other words, robust smoothing here creates a new frequency component at $\omega = 0$.

What kind of effects should we look out for? First, it is clear that only sharp and high peaks in the spectrum of X can have a noticeable and misleading leakage effect. We assert: if the spectrum of X has peaks at the frequencies $\omega_1, \omega_2, \ldots$ then the smoothed process Z may show peaks at frequencies which are sums and differences of an odd number of (not necessarily distinct) ω_i's, i.e. at frequencies

$$\begin{aligned}&\omega_i\\&\omega_{i_1} \pm \omega_{i_2} \pm \omega_{i_3}\\&\omega_{i_1} \pm \omega_{i_2} \pm \omega_{i_3} \pm \omega_{i_4} \pm \omega_{i_5}\\&\text{etc.}\end{aligned} \tag{7.1}$$

For instance, two peaks at ω_1 and ω_2 may cause leakage to $2\,\omega_1 - \omega_2$, but not to $\omega_2 - \omega_1$.

In order to demonstrate this assertion, assume that $\psi(x)$ is a continuous odd function. By the Weierstrass theorem it can be approximated arbitrarily closely on any bounded interval by a polynomial in odd powers of x. Let

$$X_t = \sum \xi_k \, e^{i\,\omega_k t} \tag{7.2}$$

be a real valued finite linear combination of trigonometric functions. If ψ is a polynomial in odd powers of X, then $\psi\,(X_t)$ has a representation

$$\psi(X_t) = \sum \eta_k \, e^{i\,\tilde{\omega}_k t} \tag{7.3}$$

with frequencies $\tilde{\omega}_k$ of the form (7.1), and the same holds for all steps of the iterations (2.21). The assertion follows now through a passage to the limit.

REFERENCES.

D. F. Andrews, P. J. Bickel, F. R. Hampel, P. J. Huber, W. H. Rogers, and J. W. Tukey (1972), Robust Estimates of Location, Princeton University Press.

R. Dutter (1977), Numerical Solution of Robust Regression Problems: Computational Aspects, a Comparison, J. of Statistical Computation and Simulation, 5, p. 207-238.

Th. Gasser and B. Krumm (1978), Personal Communication

P. J. Huber and R. Dutter (1974), Numerical Solution of Robust Regression Problems. COMPSTAT 1974, Proc. in Computational Statistics, ed. G. Bruckmann, Physica Verlag, Vienna.

R. H. Largo, W. Stützle, Th. Gasser, P. J. Huber, A. Prader (1978), Analysis of the Adolescent Growth Spurt Using Smoothing Spline Functions. To appear in Annals of Human Biology.

R. V. Lenth (1977), Robust Splines, Comm. in Statistics, A6, p. 847-854.

C. L. Mallows (1977), (Manuscript), Some Theory of Non-Linear Smoothers.

C. H. Reinsch (1967), Smoothing by spline functions, Numerische Mathematik, 10, p. 177-183.

W. Stützle (1977), Estimation and Parameterization of Growth Curves, Doctoral Dissertation, ETH Zürich.

J. W. Tukey (1977), Exploratory Data Analysis, Addison-Wesley Publishing Company.

G. Wahba and S. Wold (1975), A completely automatic french curve: Fitting spline functions by cross-validation, Comm. in Statistics, 4, p. 1-17.

Swiss Federal Institute of Technology
Zurich, Switzerland

Robust Pitman-like Estimators

M. Vernon Johns

The principal categories of estimators whose robustness properties have been studied intensively are, of course, (see, e.g., Huber 1977)

(1) L-ESTIMATORS: adaptive or non-adaptive linear combinations of order statistics,

(2) R-ESTIMATORS: related to rank order tests, and

(3) M-ESTIMATORS: analogs of maximum likelihood estimators.

Since it is customary and plausible to require that robust estimators of location, scale, and linear regression satisfy the usual invariance conditions (see, e.g., Andrews, et al. 1972), it seems reasonable to introduce a fourth class of potentially robust estimators based on the best invariant estimators, i.e,

(4) P-ESTIMATORS: analogs of Pitman estimators.

Consider the problem of estimating a simple location parameter θ. The best estimator of θ invariant under translations based on n iid observations $X_1, X_2, \ldots, X_n$ with common density $\gamma(x - \theta)$ is then the Pitman estimator

$$T_n = \frac{\int \theta \prod \gamma(X_i - \theta)d\theta}{\int \prod \gamma(X_i - \theta)d\theta} \quad . \tag{1}$$

Proceeding with a rationale much like that underlying M-estimators, we consider the estimator T_n defined in terms of a function γ which is

ISBN 0-12-438150-2

(a) not necessarily the (unknown) density of the observations, and in fact

(b) not necessarily a density function at all (e.g., γ need not be integrable so long as the integrals in the numerator and denominator of (1) exist for fixed $X_1, \ldots, X_n$).

It is evident at once that such P-estimators possess certain advantages and disadvantages when compared to M-estimators. In particular,

(a) the P-estimator does not require iterative solution methods for its computation. (Clearly an advantage.)

(b) The computation of the P-estimator may require numerical integration which could be troublesome in practice. (A disadvantage which turns out not to be fatal.)

In order to investigate the proper choice of the function γ in (1) to insure robustness, it is helpful to obtain the influence function and the asymptotic variance of the estimator T_n. To this end we must first express T_n asymptotically as a functional $T(F)$, independent of n, on the space of distribution functions. We observe that

$$T_n = \frac{\int \theta \exp\{\Sigma \log \gamma(X_i - \theta)\}d\theta}{\int \exp\{\Sigma \log \gamma(X_i - \theta)\}d\theta} , \tag{2}$$

$$= \frac{\int \theta \exp\{n \int \log \gamma(x - \theta)dF_n(x)\}d\theta}{\int \exp\{n \int \log \gamma(x - \theta)dF_n(x)\}d\theta} ,$$

where

$$F_n(x) = \frac{1}{n}(\# X_i\text{'s} \leq x) = \text{empirical c.d.f. of the sample} .$$

Replacing F_n by F, the true c.d.f. of the X_i's, in (2) does not immediately yield a functional $T(\cdot)$, independent of n. Let

$$h(t,F) = \int \log \gamma(x - t)dF(x) ,$$

$$\dot{h}(t,F) = \frac{\partial}{\partial t} h(T,F) ,$$

and

$$\ddot{h}(t,F) = \frac{\partial^2}{\partial t^2} h(t,F) \quad .$$

Then if h(t,F) has a parabolic maximum as a function of t at $t = t_F$, a well-known result from asymptotic analysis (see, e.g., de bruijn, 1961, Ch. 4) shows that as $n \to \infty$,

$$\int \theta \exp\{n \int \gamma(x-\theta)dF(x)\}d\theta$$

$$= \int t \exp\{n\,h(t,F)\}dt \sim \left(\frac{-2t_F}{\ddot{h}(t_F,F)}\right) n^{-\frac{1}{2}} \exp\{n\,h(t_F,F)\} \quad ,$$

and

$$\int \exp\{n \int \gamma(x-\theta)dF(x)\}d\theta$$

$$= \int \exp\{n\,h(t,F)\}dt \sim \left(\frac{-2}{\ddot{h}(t_F,F)}\right) n^{-\frac{1}{2}} \exp\{n\,h(t_F,F)\} \quad .$$

Hence, $T(F) = t_F$ is the required functional where T(F) is determined implicitly by the condition $\dot{h}(T(F),F) = 0$ or under appropriate regularity conditions.

$$\int \frac{\gamma'(x-T(F))}{\gamma(x-T(F))} dF(x) = 0 \quad . \tag{3}$$

Assuming now and henceforth that F is symmetric, as well as continuous, the influence function of $T(\cdot)$ satisfying (3) is defined by

$$IC(x;T,F) = \lim_{\varepsilon \to 0} \frac{1}{\varepsilon}\{T((1-\varepsilon)F + \varepsilon\,\delta_x) - T(F)\} \quad ,$$

where $\delta_x(y)$ is the c.d.f. degenerate at $y = x$. A standard argument similar to that used for M-estimators (see, e.g., Huber 1977) shows that

$$IC(x;F,T) = \frac{\frac{\partial}{\partial x} \log \gamma(x)}{\int \frac{\partial^2}{\partial x^2} \log \gamma(x)dF(x)} \quad . \tag{4}$$

The limiting normal distribution of $\sqrt{n}(T_n - \theta)$ then has variance given by

$$V(F,T) = E_F[IC(X;F,T)]^2 \quad , \tag{5}$$

$$= \frac{\int \left[\frac{\partial}{\partial x} \log \gamma(x)\right]^2 dF(x)}{\left[\int \frac{\partial^2}{\partial x^2} \log \gamma(x) dF(x)\right]^2} \quad .$$

Observe that if $\gamma(x) = f(x) = F'(x)$, where F is the common c.d.f. of the X_i's, then $V(F,T)$ = the reciprocal of the Fisher information as it should, since the Pitman estimate is then fully efficient.

Scale Invariance. Candidates for robust estimators of location should also be scale invariant. The simplest way to accomplish this is to follow the method often advocated in the M-estimator case and insert a suitable scale functional S(F) into the location estimator. This leads to P-estimators of the form

$$T_n = \frac{\int \theta \, \Pi \, \gamma\left(\frac{X_i - \theta}{S(F_n)}\right) d\theta}{\int \Pi \, \gamma\left(\frac{X_i - \theta}{S(F_n)}\right) d\theta} \quad , \tag{6}$$

where, e.g., $S(F_n) = c \cdot (MAD)$ for some $c > 0$, where (MAD) = the median absolute deviation from the median. It was found that estimators of this form performed reasonably well but could be improved significantly by using a procedure more closely related to the location and scale invariant Pitman estimator.

For observations $X_1, X_2, \ldots, X_n$ which are iid with density $\frac{1}{\sigma} \gamma(\frac{x - \theta}{\sigma})$ the location and scale invariant Pitman estimator of location may be written as

$$T_n^* = \frac{\int \frac{1}{\sigma^3} \int \theta \, \Pi \frac{1}{\sigma} \gamma\left(\frac{X_i - \theta}{\sigma}\right) d\theta d\sigma}{\int \frac{1}{\sigma^3} \int \Pi \frac{1}{\sigma} \gamma\left(\frac{X_i - \theta}{\sigma}\right) d\theta d\sigma} \tag{7}$$

$$= \int T_n(\sigma;\gamma) W_n(\sigma;\gamma) d\sigma$$

where $T_n(\sigma;\gamma)$ = the location invariant Pitman estimator given by (1) with $\gamma(x)$ replaced by $\gamma(\frac{x}{\sigma})$, and the weighting function $W_n(\sigma;\kappa)$ is given by

$$W_n(\sigma;\kappa) = \frac{\frac{1}{\sigma^3} \int \Pi \frac{1}{\sigma} \kappa\left(\frac{X_i - \theta}{\sigma}\right) d\theta}{\int \frac{1}{\sigma^3} \int \Pi \frac{1}{\sigma} \kappa\left(\frac{X_i - \theta}{\sigma}\right) d\theta d\sigma} \ . \tag{8}$$

Three remarks should be made at this point:

(i) Using a function $\kappa(\cdot)$ in (8) different from the function $\gamma(\cdot)$ might lead to improved performance for T_n^*.

(ii) The double integrals appearing in (7) and (8) create practical difficulties.

(iii) $W_n(\sigma;\kappa)$ is a random density function which, as n increases, tends to concentrate at σ = the Pitman estimate for σ when θ is unknown (using κ as the "density" for the observations).

This suggests using the P-estimator T_n given by (6) with $S(F_n) = S_n$ given by

$$S_n = \frac{\int \frac{1}{\sigma^2} \Pi \frac{1}{\sigma} \kappa\left(\frac{X_i - M}{\sigma}\right) d\sigma}{\int \frac{1}{\sigma^3} \Pi \frac{1}{\sigma} \kappa\left(\frac{X_i - M}{\sigma}\right) d\sigma} \ , \tag{9}$$

where M = a consistent estimator of the center of symmetry such as the sample median. Here S_n approximates the Pitman estimator of scale.

Approximations for Integrals. If the P-estimators are not to be computationally too complex for practical use, the integrals in (6) and (9) must be approximated rather crudely. The following simple approximations were found to work well:

$$\hat{T}_n = \frac{\sum_j b_j t_j \prod_i \gamma\left(\frac{X_i - t_j}{\hat{S}_n}\right)}{\sum_j b_j \prod_i \gamma\left(\frac{X_i - t_j}{\hat{S}_n}\right)} , \tag{10}$$

where the t_j's are equally spaced on $(M - 4\hat{S}_n, M + 4\hat{S}_n)$, the b_j's are the appropriate coefficients for Simpson's rule, and

$$\hat{S}_n = c_1 \cdot \frac{\sum_j \frac{b_j}{s_j^2} \prod_i \frac{1}{s_j} \kappa\left(\frac{X_i - M}{s_j}\right)}{\sum_j \frac{b_j}{s_j^3} \prod_i \frac{1}{s_j} \kappa\left(\frac{X_i - M}{s_j}\right)} , \tag{11}$$

where the s_j's are equally spaced on $(0, 5 \cdot (\mathrm{MAD}))$, the b_j's are the appropriate coefficients for Simpson's rule, and c_1 is a constant to be chosen empirically.

The numerator and denominator of (10) are the Simpson's rule approximations to the corresponding integrals of (6). Similarly, the sums in (11) approximate the integrals of (9).

The Choice of $\gamma(\cdot)$ and $\kappa(\cdot)$. A systematic search for optimally robust choices for $\gamma(\cdot)$ and $\kappa(\cdot)$ has not yet been implemented, but certain ad hoc choices have been investigated empirically by means of simulations. In particular, γ's of the form

$$\gamma(x) = \frac{1}{1 + x^2} + \frac{a + bx^2}{(1 + x^2)^3} ,$$

for various choices of a and b, have been used. The rationale for the choice of the functional form was to approximate the normal density in the central region and to provide a slow approach to zero in the tails. Two pairs of values for a and b appearing in $\gamma(\cdot)$, namely, $(a,b) = (0,.5)$ and $(a,b) = (1.5,2.5)$, have been found to yield interesting results. A comparison of $\gamma(x)$ with the standard normal density is shown in Figure 1.

The $\kappa(\cdot)$ functions which so far have been found to give reasonably good results are of the following peculiar discontinuous form:

$$\kappa(x) = \begin{cases} \dfrac{1}{1 + (|x| - c)^2} , & 0 \leq |x| \leq d \\ \dfrac{1}{g|x|} , & d \leq |x| \leq h \\ \dfrac{h}{gx^2} , & |x| > h \end{cases} .$$

Typical values of the parameters are $c = .7$, $d = 3$, $g = 10$, $h = 100$.

Simulation Results. To evaluate the performance of P-estimators it was decided to compare them with the one-step "bisquare" M-estimator advocated by Beaton and Tukey (1974) and Gross (1976). This one-step M-estimator (a w-estimator in Tukey's terminology) is given by

$$\tilde{T}_n = M + \frac{\tilde{S}_n \sum_i \psi_1\left(\dfrac{X_i - M}{\tilde{S}_n}\right)}{\sum_i \psi_2\left(\dfrac{X_i - M}{\tilde{S}_n}\right)} , \qquad (12)$$

where

$$\psi_1(x) = \begin{cases} x(1 - x^2)^2 , & |x| \leq 1 \\ 0 , & |x| > 1 , \end{cases}$$

and

$$\psi_2(x) = \begin{cases} (1 - x^2)(1 - 5x^2) , & |x| \leq 1 \\ 0 , & |x| > 1 , \end{cases}$$

$\tilde{S}_n = c_2 \cdot (\text{MAD})$ and M again represents the sample median. Figure 2 shows the influence curves for the bisquare estimator and the P-estimator using $\gamma(x)$.

The choice of distributions used in the simulation studies was guided by Tukey's concept of "triefficiency" (Beaton and Tukey 1974). Thus, the distributions used to compare efficiencies were:

1. Standard Normal Distribution (N(0,1).
2. Normal Over Uniform Distribution (N/U) (i.e., the distribution of a standard normal variate divided by an independent uniform —(0,1) variate).

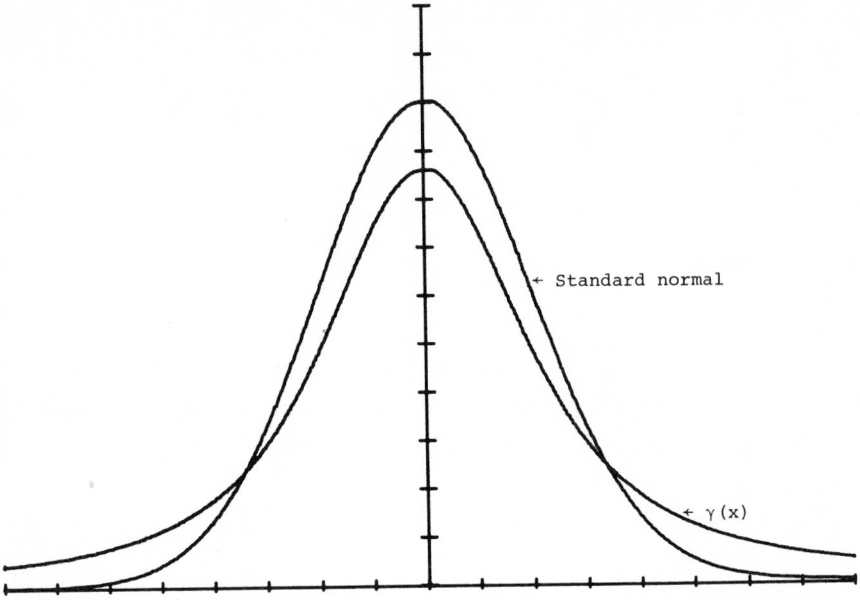

Fig. 1. Comparison of the shapes of $\gamma(x)$ and the standard normal density.

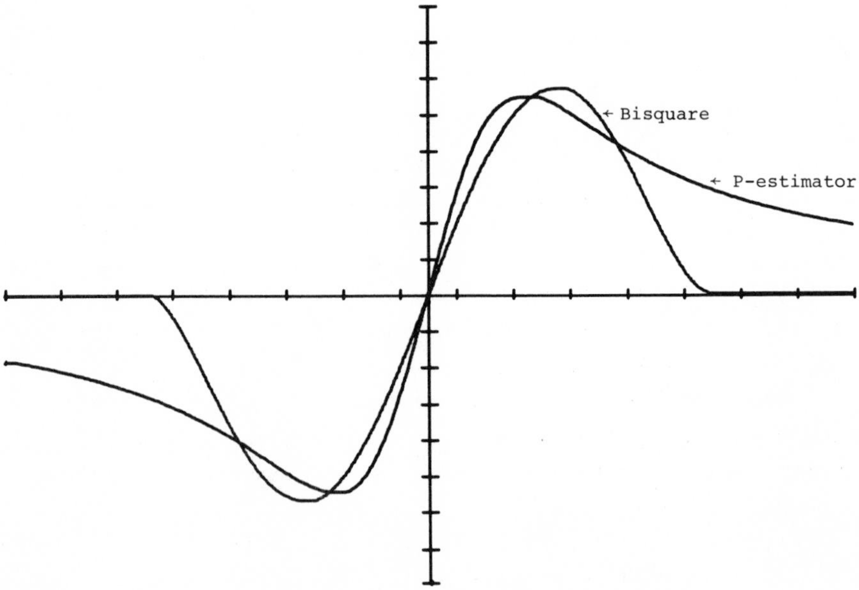

Fig. 2. Comparison of influence curve shapes for the bisquare and the $\gamma(x)$ P-estimator.

3. "One Wild" Normal Distribution (1WN) (i.e., a sample of size 20 consists of 19 standard normal variates and one normal variate with variance = 100).

The sample size used throughout is n = 20.

Some typical simulation results are summarized in Tables 1 and 2. Here V_M = variance of the one-step "bisquare" M-estimator, V_P = variance of the P-estimator and V_O = variance of the "best" estimator for each particular distribution as given by Andrews, et al. (1972). The "Princeton swindle" for variance reduction (Andrews et al. 1972) was used throughout the present simulation study. In interpreting the results, it should be borne in mind that because of the high correlations between the estimators involved, the efficiency ratios are much more stable and reliable than are the estimated efficiencies themselves. It was found that much larger simulation sizes were required to obtain reasonably stable results for the long-tailed (N/U) distribution than were needed for the normal case.

TABLE 1

Sample Size n = 20. $c_1 = 1.2$, $c_2 = 6.4$

γ: $a = 1.5$, $b = 2.5$.

κ: $c = .6$, $d = 3$, $g = 10$, $h = 100$.

Distribution	Number of Samples	Efficiency Ratio (V_M/V_P)	Efficiency of P-Est. (V_O/V_P)
N(0,1)	2,000	1.04	.92
N/U	10,000	1.04	.92
1WN	2,000	1.00	.95

This table, in particular, shows that the performance of even a very good M-estimator such as the bisquare may be appreciably improved upon by a suitable P-estimator.

TABLE 2†

Sample Size n = 20 γ: a = 0, b = 1 Scale Estimator $S_n = 2.15 \times$ (MAD)			
Distribution	Number of Samples	Efficiency Ratio (V_M/V_P)	Efficiency of P-Est. (V_O/V_P)
N(0,1)	2,000	1.02	.91
N/U	10,000	1.00	.89
1WN	2,000	.98	.94

†Using trapezoidal approximations for the integrals in the P-estimator.

The Regression Case (conjectures). For the linear regression model

$$X = C\beta + \varepsilon \quad ,$$

where X is the vector of n observations, β is the p-vector of regression coefficients, C is the $n \times p$ design matrix, and ε is the vector of n iid errors with common symmetric density $\gamma(\cdot)$. The location and scale invariant Pitman estimators for the components of β are given by

$$\hat{\beta}_i = \frac{\int \frac{\beta_i}{\sigma^{n+3}} \prod_j \gamma\left(\frac{X_j - (C\beta)_j}{\sigma}\right) d\beta_1 \ldots d\beta_p \, d\sigma}{\int \frac{1}{\sigma^{n+3}} \prod_j \gamma\left(\frac{X_j - (C\beta)_j}{\sigma}\right) d\beta_1 \ldots d\beta_p \, d\sigma} \tag{13}$$

for i = 1,2,...,p.

The following remarks and conjectures are perhaps relevant:

(i) It seems likely that for choices of $\gamma(\cdot)$ which lead to robust estimates for location, the estimates $\hat{\beta}_i$ would also be robust.

(ii) The presence of (p + 1)-fold integrals in (13) presents a difficult computational problem in practice. The problem may not be hopeless, however, in view of recent advances in techniques for numerical multiple integration (see, e.g., Friedman 1977).

(iii) If good initial estimates $\tilde{\beta}_i$ were available, it is possible that the following strategy would give satisfactory results:

(a) Estimate σ using a single ratio of approximate integrals as in the location and scale case (substituting $\tilde{\beta}$ for β).

(b) Substitute in (13) the estimate of σ and the components of $\tilde{\beta}$ for all β's except β_i. Integrate only with respect to β_i and approximate the integrals as in the location case.

The components of $\tilde{\beta}$ could be replaced in (b) by the improved estimates of the earlier β_i's as they become available. Steps (a) and (b) could be iterated.

The generalization of the P-estimator concept to the regression case is under continuing investigation.

REFERENCES

1. Andrews, D. F., Bickel, P. J., Hampel, F. R., Huber, P. J., Rogers, W. H., and Tukey, J. W. (1972). Robust Estimates of Location: Survey and Advances. Princeton University Press.

2. Beaton, A. E. and Tukey, J. W. (1974). The fitting of power series, meaning polynomials, illustrated on band-spectroscopic data. Technometrics, 16, 147-185.

3. de Bruijn, N. G. (1961). Asymptotic Methods in Analysis, 2nd ed. Amsterdam: North-Holland Publishing Co.

4. Friedman, J. H. (1977). A nested partitioning algorithm for numerical multiple integration. Technical Report SLAC PUB-2006, Stanford University.

5. Gross, A. M. (1976). Confidence interval robustness with long-tailed symmetric distributions. JASA, 71, 409-416.

6. Huber, P. J. (1977). Robust Statistical Procedures. SIAM Regional Conference Series in Applied Mathematics No. 27.

This research was supported by U.S. Army Research Office Contract No. DAAG29-76-G-0213.

Department of Statistics
Stanford University
Stanford, California 94305

Robust Estimation in the Presence of Outliers

H. A. David

1. INTRODUCTION.

In this paper we shall examine the bias and mean square error of various estimators of location and scale in the presence of an unidentified outlier. The estimators L_n (L-estimators) studied are linear functions of the order statistics, viz.

$$L_n = \sum_{i=1}^{n} a_{in} Z_{i:n}, \tag{1}$$

where the a_{in} are constant coefficients and the $Z_{i:n}$ $(i = 1, 2, \ldots, n)$ are the original observations Z_i, arranged in ascending order:

$$Z_{1:n} \le Z_{2:n} \le \cdots \le Z_{n:n} .$$

Familiar examples are median, trimmed means, and Winsorized means as estimators of location, and quasi-ranges as estimators of scale.

Our main statistical model, the k-outlier model, divides the Z_i into n-k X's and k Y's as follows:

$$\begin{aligned} &X_1, X_2, \ldots, X_{n-k} \text{ have cdf } F(x) \text{ and pdf } f(x), \\ &Y_1, Y_2, \ldots, Y_k \quad \text{have cdf } G(x) \text{ and pdf } g(x). \end{aligned} \tag{2}$$

ISBN 0-12-438150-2

Also $X_1, \ldots, X_{n-k}, Y_1, \ldots, Y_k$ are mutually independent. Of course, it is not known whether a particular Z is an X or a Y variate. We concentrate on the case k = 1. However, we shall make some comparisons in Section 5 of the k-outlier model with the mixture model, according to which r.v.'s Z_i' are a random sample with cdf $H_\gamma(x)$ given by

$$H_\gamma(x) = (1-\gamma) F(x) + \gamma G(x), \tag{3}$$

where γ is a constant $(0 \leq \gamma \leq 1)$ which is usually small (say $\gamma \leq 0.2$); see Gastwirth and Cohen [8].

More specifically, we shall study the following special cases of (2) corresponding to (a) a change in location and (b) a change in scale:

$$X_j \ (j = 1, 2, \ldots, n-1) \text{ has cdf } F(x) \quad \begin{array}{ll} Y \text{ has cdf } F(x-\lambda) & \text{(4a)} \\ Y \text{ has cdf } F(x/\tau), & \text{(4b)} \end{array}$$

where $\tau > 0$. Equivalently Y may be expressed as $X_n + \lambda$ in (4a) and as τX_n in (4b). The numerical results to be presented are confined to $F(x) = \Phi(x)$, the standard normal cdf.

All these models are well known and go back at least to Dixon [7]. Models (4) are often termed "slippage models".

2. SOME REMARKS ON THE RELATION BETWEEN ROBUST ESTIMATION AND TESTS FOR OUTLIERS.

Publications on robust estimation sometimes convey the impression that tests for outliers are irrelevant. The argument seems to be that robust estimators are constructed to perform reasonably well as long as the number of outliers is not too large. After all, the more extreme observations in the sample are typically given little or no weight in the robust estimator. Moreover, the argument proceeds, those observations which are rejected by some standard test for outliers may not be outliers at all; rather, we may well be dealing with a long-tailed distribution.

All this is true and would be compelling if the only purpose of tests for outliers were the cleansing of data prior to estimation of the parameters of interest. It is important, therefore, to reiterate another aim of outlier tests, crucial in the proper treatment of data, namely the identification of observations deserving closer scrutiny.

True, such scrutiny is not always practicable and may reveal only that something trivial went wrong in isolated instances. On the other hand, we may learn that the data-generating process broke down for part of the sample, indicating the need for remedial action. Or we may be able to pinpoint observations which are of special interest just because they are outlying: they may represent a hitherto unsuspected phenomenon. Of course, after scrutiny we may find nothing alarming or special about the more extreme observations. In this case the evidence points us to an underlying long-tailed distribution.

I do not want to overstate the case for tests of outliers. They are often lacking in power, a point to which we shall return later. Suffice it to say that outlier tests applied at the usual levels of significance do not obviate the need for robust estimators. It is sometimes suggested that outlier tests be performed at the 10% or even 20% level. Raising the significance level will certainly improve the robustness of estimators based on the surviving observations. However, (a) there are more effective ways of constructing robust estimators and (b) too many observations will tend to be identified for further scrutiny.

See also the soon-to-be published Wiley text Outliers in Statistical Data by Barnett and Lewis [2], especially Chapter 4.

3. FIRST TWO MOMENTS OF LINEAR FUNCTIONS OF ORDER STATISTICS IN THE PRESENCE OF A SINGLE OUTLIER

Let $Z_{r:n}(\lambda)$ and $Z^*_{r:n}(\tau)$ denote the r-th order statistic ($r = 1, 2, \ldots, n$) in (4a) and (4b), respectively. Further, let

$$\mu_{r:n}(\lambda) = E\,Z_{r:n}(\lambda),$$

$$\sigma_{rs:n}(\lambda) = \operatorname{cov}(Z_{r:n}(\lambda),\ Z_{s:n}(\lambda)), \quad s = 1, 2, \ldots, n,$$

with corresponding meanings for $\mu^*_{r:n}(\tau)$, $\sigma^*_{rs:n}(\tau)$. Of course, these moments depend also on $F(x)$ and can generally be evaluated only numerically. For $F(x) = \Phi(x)$ see David et al. [4]. However, for any F possessing a second moment we have the obvious results

$$\mu_{r:n}(0) = \mu^*_{r:n}(1) = EX_{r:n} \equiv \mu_{r:n},$$
$$\sigma_{rs:n}(0) = \sigma^*_{rs:n}(1) = \operatorname{cov}(X_{r:n}, X_{s:n}) \equiv \sigma_{rs:n}.$$

Also under mild regularity conditions on F, the following intuitively plausible results may be established (David and Shu [6]):

$$\mu_{r:n}(\infty) = \mu_{r:n-1} \qquad r = 1, 2, \ldots, n-1 \tag{5}$$

$$\mu_{r:n}(-\infty) = \mu_{r-1:n-1} \qquad r = 2, 3, \ldots, n \tag{6}$$

$$\mu_{1:n}(-\infty) = -\infty, \quad \mu_{n:n}(\infty) = \infty \tag{7}$$

$$\sigma_{rs:n}(\infty) = \sigma_{rs:n-1} \qquad r, s = 1, 2, \ldots, n-1 \tag{8}$$

$$\sigma_{rs:n}(-\infty) = \sigma_{r-1, s-1:n-1} \qquad r, s = 2, 3, \ldots, n \tag{9}$$

and if $F(0) = \frac{1}{2}$

$$\mu^*_{r:n}(\pm\infty) = \tfrac{1}{2}(\mu_{r-1:n-1} + \mu_{r:n-1}) \qquad r = 2, 3, \ldots, n-1 \tag{10}$$

$$\sigma^*_{rs:n}(\pm\infty) = \tfrac{1}{2}(\sigma_{r-1, s-1:n-1} + \sigma_{rs:n-1}) \qquad r, s = 2, 3, \ldots, n-1 \tag{11}$$

$$\mu^*_{1:n}(\pm\infty) = -\infty, \quad \mu^*_{n:n}(\pm\infty) = \infty. \tag{12}$$

We now take $F(x)$ to be the cdf of a symmetric distribution for which we wish to estimate parameters of location and scale by linear functions of order statistics. It is natural in this situation to confine consideration to the class of location estimators

$$M(\lambda) = \sum_{i=1}^{n} a_i Z_{i:n}(\lambda) \qquad a_i = a_{n+1-i} \geq 0, \quad \sum_{i=1}^{n} a_i = 1 \tag{13}$$

and to the class of scale estimators

$$D(\lambda) = \sum_{i=1}^{n} b_i Z_{i:n}(\lambda) \qquad b_i = -b_{n+1-i} \geq 0 \text{ for } i > [\tfrac{1}{2}n]. \tag{14}$$

These are written here for case (4a). Corresponding expressions apply for (4b). The dependence of estimators and coefficients on n, made explicit in (1), is here understood. For studying the properties of these estimators it is convenient to take $F(x)$ to be standardized with zero mean and unit variance. The first two moments of $M(\lambda)$ are discussed in [6]. We now concentrate on $D(\lambda)$, where the b_i are chosen so that for $\lambda = 0$, $D(0)$ is an unbiased estimator of the desired scale parameter. From the symmetry of F about zero we see that in (4a) (with $\sim$ denoting "is distributed as")

$$(X_1, \ldots, X_{n-1}, X_n - \lambda) \sim (-X_1, \ldots, -X_{n-1}, -X_n - \lambda)$$

so that

$$Z_{r:n}(-\lambda) \sim -Z_{n+1-r:n}(\lambda), \tag{15}$$

$$\mu_{r:n}(-\lambda) = -\mu_{n+1-r:n}(\lambda), \tag{16}$$

$$\sigma_{rs:n}(-\lambda) = \sigma_{n+1-s,n+1-r:n}(\lambda). \tag{17}$$

Then by (14) we have

$$ED(-\lambda) = ED(\lambda), \quad ED^2(-\lambda) = ED^2(\lambda). \tag{18}$$

Also, for $i > [\frac{1}{2}n]$ and $\lambda > 0$, it is clear that $Z_{i:n}(\lambda) - Z_{n+1-i:n}(\lambda)$ is stochastically increasing in λ. Since $D(\lambda)$ may be written as

$$D(\lambda) = \sum_{i=[\frac{1}{2}n]+1}^{n} b_i[Z_{i:n}(\lambda) - Z_{n+1-i:n}(\lambda)],$$

it and $D^2(\lambda)$ are also stochastically increasing in λ. Thus $ED(\lambda)$ and $ED^2(\lambda)$ are monotonically increasing in $|\lambda|$. In view of (7), $D(\pm\infty)$ has finite first two moments only if $b_1 = -b_n = 0$ in which case

$$ED(\pm\infty) = E\lim_{\lambda\to\infty} D(\lambda)$$

$$= \lim_{\lambda\to\infty} ED(\lambda) \text{ by the monotone convergence theorem.}$$

Hence by (5) we have

$$ED(\pm\infty) = \sum_{i=2}^{n-1} b_i \mu_{i:n-1}\,. \tag{19}$$

Likewise we have

$$ED^2(\pm\infty) = [ED(\pm\infty)]^2 + \sum_{i=2}^{n-1}\sum_{j=2}^{n-1} b_i b_j \sigma_{ij:n-1}\,. \tag{20}$$

Writing

$$D^*(\tau) = \sum_{i=2}^{n-1} b_i Z^*_{i:n}(\tau) \tag{21}$$

one can easily show (cf. [6])

$$ED^*(\infty) = ED(\infty) \text{ and } ED^{*2}(\infty) = ED^2(\infty). \tag{22}$$

4. NUMERICAL RESULTS IN THE NORMAL CASE FOR ESTIMATORS OF σ

The following estimators will be considered under model (4a) with $F = \Phi$:

BL = Best linear unbiased estimator (BLUE) of σ

G = A. K. Gupta's [10] estimator

BL(1) = BL for 1 observation removed at each end

G(1) = G for 1 observation removed at each end

BL(2) = BL for 2 observations removed at each end

G(2) = G for 2 observations removed at each end

GN = Gini's mean difference

$$= \text{const.} \sum_{i=1}^{n} \left(i - \frac{n+1}{2}\right) Z_{i:n}$$

$W = Z_{n:n} - Z_{1:n}$ (range)

$W(1) = Z_{n-1:n} - Z_{2:n}$ (first quasi-range)

$W(2) = Z_{n-2:n} - Z_{3:n}$ (second quasi-range)

$J_2 = R + R(1)$ (thickened range)

$J_3 = R + R(1) + R(2)$ (thickened range).

For a general account of these estimators see e.g. David [3]. Divisors were applied where necessary to render the estimators unbiased in the null case ($\lambda = 0$). Bias and mean square error were obtained as functions of λ for $n = 10$, 20 and for $n = 5$ where applicable. Since the results have the same features for the three sample sizes we confine ourselves to $n = 10$. Table 1 gives the coefficients b_i of $D(\lambda)$ in this case, those for BL, BL(1), BL(2) being taken from Sarhan and Greenberg [11], pp. 222-23.

Table 1. Coefficients b_i for Various Unbiased Estimators $D(\lambda)$ of σ

	b_6	b_7	b_8	b_9	b_{10}
BL	.0142	.0436	.0763	.1172	.2044
G	.0155	.0475	.0829	.1265	.1944
BL(1)	.0201	.0616	.1074	.4034	0
G(1)	.0386	.1182	.2064	.3150	0
BL(2)	.0310	.0947	.7021	0	0
G(2)	.1046	.3202	.5592	0	0
GN	.0197	.0591	.0985	.1379	.1772
W	0	0	0	0	.3249
W(1)	0	0	0	.4993	0
W(2)	0	0	.7621	0	0
J_2	0	0	0	.1968	.1968
J_3	0	0	.1564	.1564	.1564

Table 2. Bias and MSE of the Estimators in Table 1 for Sample Size n = 10 When One Observation is from a N(λ, 1) Population and the Others from N(0, 1)

(a) Bias

λ	0	0.5	1	1.5	2	3	4	∞
BL	0.0001	.0125	.0488	.1062	.1805	.3613	.5607	∞
G	0	.0124	.0485	.1050	.1775	.3517	.5420	∞
BL(1)	0	.0122	.0443	.0861	.1261	.1765	.1917	.1944
G(1)	0	.0121	.0438	.0842	.1219	.1674	.1804	.1826
BL(2)	0	.0120	.0423	.0783	.1086	.1385	.1443	.1448
G(2)	0	.0120	.0421	.0775	.1069	.1352	.1404	.1409
GN	0	.0124	.0480	.1030	.1723	.3348	.5094	∞
W	0	.0126	.0516	.1190	.2149	.4754	.7847	∞
W(1)	0	.0122	.0448	.0880	.1306	.1862	.2038	.2070
W(2)	0	.0120	.0424	.0787	.1093	.1399	.1459	.1464
J_2	0	.0124	.0490	.1068	.1816	.3614	.5557	∞
J_3	0	.0124	.0476	.1010	.1668	.3159	.4716	∞

(b) MSE

λ	0	0.5	1	1.5	2	3	4	∞
BL	.0576	.0592	.0652	.0792	.1056	.2104	.3965	∞
G	.0577	.0592	.0651	.0786	.1037	.2017	.3736	∞
BL(1)	.0824	.0845	.0910	.1016	.1149	.1386	.1494	.1521
G(1)	.0849	.0870	.0934	.1033	.1152	.1351	.1435	.1456
BL(2)	.1292	.1324	.1410	.1529	.1648	.1799	.1840	.1845
G(2)	.1341	.1374	.1462	.1579	.1692	.1830	.1865	.1870
GN	.0581	.0597	.0654	.0781	.1010	.1877	.3361	∞
W	.0671	.0690	.0774	.0996	.1462	.3537	.7561	∞
W(1)	.0853	.0875	.0945	.1064	.1219	.1509	.1646	.1682
W(2)	.1301	.1333	.1421	.1542	.1665	.1824	.1868	.1873
J_2	.0592	.0608	.0670	.0814	.1083	.2123	.3921	∞
J_3	.0594	.0610	.0666	.0786	.0995	.1743	.2974	∞

Apart from noting the obvious dichotomy into estimators placing finite weight and those placing zero weight on the extremes we may comment on the following features of Table 2:

(i) The dilemma of scale estimators is that for good efficiency in the null case considerable weight must be placed on the more extreme observations. Consequently the use of a trimmed sample results in a much greater increase in MSE for $\lambda = 0$ than in the estimation of location (cf. Table 2 of [6]).

(ii) If bias and MSE are plotted for each estimator as functions of λ there will tend to be a pulling apart of the bias curves as λ increases but often a crossover for the MSE curves.

(iii) The simpler Gupta estimators are slightly superior to the corresponding BLUE's in bias, as may be expected from their placing less weight on the extremes. The MSE comparisons are more complicated.

(iv) W(1) outperforms R in MSE for $\lambda \geq 2$ and is uniformly but not greatly inferior to both BL(1) and G(1). It is uniformly superior to W(2).

(v) Those estimators having finite bias and MSE for $\lambda = \infty$ have reached nearly maximal bias and MSE for $\lambda = 3$ and very nearly for $\lambda = 4$. Similar results hold for location estimators (See Tables 1 and 2 of [6]).

(vi) Numerical results for model (4b) are similar. This is perhaps not too surprising in view of (22). Since, for $\tau > 1$, model (4b) results when λ in (4a) is given a normal $N(0, \tau^2-1)$ distribution (with λ independent of $X_1, X_2, \ldots, X_n$) we have also for $j = 1, 2, \ldots$

$$D^{*j}(\tau) = \int_{-\infty}^{\infty} D^j(\lambda) \frac{\exp - \frac{\lambda^2}{2(\tau^2-1)}}{\sqrt{2\pi(\tau^2-1)}} d\lambda .$$

We come now to an important question equally relevant for location estimators (cf. Tukey [12]): Would an observation corresponding to a large value of λ have much of a chance of surviving a reasonable outlier test or even inspection of the data? Let us examine this question in circumstances most favorable to detection, namely when the idealized model (4a) holds and we know $\lambda \geq 0$. In this case the optimal procedure, is to declare the largest observation an outlier if the internally studentized extreme deviate

$$\frac{Z_{n:n} - \overline{Z}}{S} \quad \text{exceeds } c_\alpha,$$

where c_α is tabulated by Grubbs [9]. The following probabilities of detection are available from David and Paulson [5]:

α λ	1	2	3	4	5	6
.05	.035	.149	.382	.667	.874	
.01		.046	.155	.358	.607	.814

It is clear that even in this idealized situation an outlier may easily be small enough to escape detection but large enough to cause a nearly maximal increase in bias or MSE! In spite of this lack of power of outlier tests, alluded to in Section 2, such tests seem to me worthwhile, if when we are successful in detection we can learn something important about the data.

5. SOME RELATIONS BETWEEN THE k-OUTLIER MODEL AND THE MIXTURE MODEL

We return now to a comparison of models (2) and (3) introduced in Section 1. The cdf $J(x|k)$, say, of a randomly chosen Z_i $(i = 1, 2, \ldots, n)$ in (2) may clearly be written as

$$J(x|k) = (1 - \frac{k}{n})F(x) + \frac{k}{n} G(x). \tag{23}$$

If k is given a binomial $b(\gamma, n)$ distribution the resulting cdf is $H_\gamma(x)$ of (3), as is well known. Directly from (23) we have also

$$J(x|n\gamma) = H_\gamma(x) \tag{24}$$

and, in obvious notation, for $k = n\gamma$ we find

$$\mu_H = \mu_J = (1-\gamma)\mu_F + \gamma\mu_G, \tag{25}$$

$$\sigma_H^2 = \sigma_J^2 = (1-\gamma)\sigma_F^2 + \gamma\sigma_G^2 + \gamma(1-\gamma)(\mu_F - \mu_G)^2. \tag{26}$$

By (24) the marginal distributions of Z_i and Z_i' (as defined prior to Eq. (3)) are identical for $k = n\gamma$. However, the joint distribution of k Z's $(k > 1)$ is not the same as that of k Z''s. In particular, with $\overline{X}$ denoting the mean of the $n-k$ X's and $\overline{Y}$ the mean of the k Y's, we have

$$\begin{aligned} \operatorname{var} \overline{Z} &= \operatorname{var} [(1-\gamma)\overline{X} + \gamma\overline{Y}] \\ &= \frac{1}{n}[(1-\gamma)\sigma_F^2 + \gamma\sigma_G^2] \\ &\le \operatorname{var} \overline{Z}' \end{aligned} \tag{27}$$

with equality (for $0 < \gamma < 1$) if $\mu_F = \mu_G$, by (26). Equality holds for model (4b) but even in this case we have the following interesting result:

If $T = t(\underset{\sim}{Z})$, $T' = t(\underset{\sim}{Z}')$ are well-behaved robust location estimators, then under the models (4)

$$\operatorname{var} T < \operatorname{var} T'.$$

Proof. Since T is robust, the ratio $\operatorname{var} T/\operatorname{var} \overline{Z}$ regarded as a function of k must have maxima for $k = 0$ and n, corresponding to pure samples, and a minimum in between. By (27) $\operatorname{var} \overline{Z}$ is a linear function of k so that $\operatorname{var} T$ must be convex in k, say

$$\operatorname{var} T = g(k) \qquad g \text{ convex}.$$

In the formula

$$\text{var } Y = \text{var } (EY|X) + E(\text{var } Y|X)$$

taking $Y = T'$, $X = K \sim b(\gamma, n)$ and noting that $T'|k = T$ and $ET = \text{const.}$, we have

$$\begin{aligned} \text{var } T' &= Eg(K) \\ &> g(EK) \quad \text{by Jensen's inequality} \\ &= \text{var } T. \end{aligned}$$

Further relations between the models are examined in a Ph. D. thesis by V. S. Shu.

6. CONCLUDING REMARKS AND SUMMARY

Together with its companion paper [6] this article represents a study of the robustness of linear functions of order statistics (L-estimators) as location or scale estimators when an outlier is present. The underlying model (4) allows for slippage in either mean or variance. In the present paper the main emphasis is on scale estimators. With the help of special tables values of bias and mean square error are tabulated for 12 estimators in the mean slippage case under normality.

It is obviously important to obtain results also for estimators not in the class of L-estimators. This is currently being done for location estimators by V. S. Shu using mainly Monte Carlo methods. The most popular robust scale estimator outside the class is the median deviation corrected for bias in normal samples: $\text{med}|Z_i - \text{med } Z_i|/0.6745$. Such Monte Carlo studies are facilitated by the exact results available for L-estimators in the slippage cases under normality.

The paper also contains comments on the relation of (a) outlier tests and robust estimation and (b) the k-outlier model and the mixture model.

REFERENCES

1. Andrews, D. F., Bickel, P. J., Hampel, F. R., Huber, P. J., Rogers, W. H., and Tukey, J. W. (1972). Robust Estimates of Location. Princeton University Press.
2. Barnett, V. and Lewis, T. (1978). Outliers in Statistical Data. Wiley, Chichester (England) and New York.
3. David, H. A. (1970). Order Statistics. Wiley, New York.
4. __________, Kennedy, W. J., and Knight, R. D. (1977). Selected Tables in Mathematical Statistics, 5, 75-204, American Mathematical Society, Providence, R. I.
5. __________ and Paulson, A. S. (1965). The performance of several tests for outliers. Biometrika, 52, 429-436.
6. __________ and Shu, V. S. (1978). Robustness of location estimators in the presence of an outlier. In: David, H. A. (Ed.), Contributions to Survey Sampling and Applied Statistics, 235-250, Academic Press, New York.
7. Dixon, W. J. (1953). Processing data for outliers. Biometrics, 9, 74-89.
8. Gastwirth, J. L. and Cohen, M. L. (1970). Small sample behavior of robust linear estimators of location. J. Amer. Statist. Assoc., 65, 946-973.
9. Grubbs, F. E. (1950). Sample criteria for testing outlying observations. Ann. Math. Statist., 21, 27-58.
10. Gupta, A. K. (1952). Estimation of the mean and standard deviation of a normal population from a censored sample. Biometrika, 39, 260-273.
11. Sarhan, A. E. and Greenberg, B. G. (Eds.) (1962). Contributions to Order Statistics. Wiley, New York.
12. Tukey, J. W. (1960). A survey of sampling from contaminated distributions. In: Olkin, I. et al. (Eds.), Contributions to Probability and Statistics, 448-485, Stanford University Press.

This work was supported by the Army Research Office. I am grateful to H. N. Nagaraja for the computation of the tables.

Statistical Laboratory
Iowa State University
Ames, Iowa 50011

Study of Robustness by Simulation: Particularly Improvement by Adjustment and Combination

John W. Tukey

1. Introduction. This account is directed to the tool-forger, to the developer and evaluator of new estimating procedures. (The following note is directed toward the user of statistical procedures.) I have attempted to emphasize principles and downplay details, except where these seem likely to help by fixing ideas. Although directed to a professional audience, its purpose is largely expository.

Some of the main points it stresses are:

1) that Monte Carlo (or experimental sampling) study should be an active process of data analysis, and not just a passive evaluation of estimate performance,

2) that, in addition to whatever leads more subtly to insight, there are formalizable procedures for learning from the details of one Monte Carlo (or experimental sampling) experiment, learning things that may provide better estimates, may suggest new candidate estimates, or may do both,

3) what some of these procedures are.

The rapid progress in providing more polyefficient and more versatile robust procedures that will also be easier to use and describe is too important to be left to

1) suggestions from asymptotic theorems, AND

2) such insights as Frank Hampel's about the need for redescending ψ-functions in centering

ISBN 0-12-438150-2

no matter how useful these two approaches are. If we can add a useful third string to our bow, we should do so.

The procedures discussed below are largely developments from, or formalizations of, appropriate aspects of what has been growing up in Princeton during the years since the appearance of the so-called Princeton Robustness Study (Andrews, et al, 1972).

For those not used to working with several alternate distributions (more generally with several alternate situations), it is important to be warned about how much change in habits of thought will be called for. Many are not used to having each estimate having 3 (or 7, or 29) distinct variances, all of which may be crucially important.

I. THE UNDERLYING STRUCTURE

2. The ideal format. The ideal picture combines definite (non-stochastic) processes

$$\{\text{data sets}\} \rightarrow \text{estimate}$$

with stochastic (random) processes

$$\text{distribution (or more general situation)} \rightarrow \{\text{data sets}\}$$

and depends for evaluation on one or more criteria, such as:

- variance,
- bias,
- mean-square error,
- mean-square deviation.

Thus, for each criterion we ideally bring to our evaluation a two-way table, say with situations labelling columns and estimates (computational processes) labelling rows.

The ideal problems are:

1) to expand our bouquet of estimates by describing new computing processes with, hopefully, desirable properties

2) to select from the current bouquet one or more favored estimates, desirable in terms of performance, simplicity of computation, and simplicity of description.

Easy to state, but likely to be hard to realize -- if for no other reason because we ordinarily do not know the precise values of the criteria for our estimates.

3. The actual format. The actual format can be thought of as another two-way table -- or group of tables -- in which

- the columns are labelled by finite approximations to situations (often lists of inner-configured data sets),
- the rows are labelled by computer-program realizations approximating (usually quite closely enough) each of a finite bouquet of carefully described estimates (calculating processes),
- the entries tell us how, in terms of one criterion, the corresponding finitely-programmed process behaved for each of the collections of data sets that are involved in the finite approximation to the corresponding situation.

We have to word the last point carefully, since even if the situation is only a distribution so that we might take just, say, 1000 samples "from" that distribution as our finite approximation to the distribution, efficient operation (Monte Carlo, not mere experimental sampling) calls for doing something a little more complicated than drawing random samples.

* the real problems *

The real problems are:

1) to expand our bouquet of estimates by describing new computing processes and then applying them to the chosen finite approximations to the chosen situations,

2) to select from the current bouquet of estimates one or more favored estimates, desirable in terms of performance, simplicity of computation, and simplicity of description,

3) to determine whether, or not, the finite approximations to the situations are close enough to the situations themselves so that any inference (not formally statistical) that we inevitably make from "performance where tested" to "performance throughout a wide range of situations encompassed by the situations tested" is essentially as sound for the finite approximations used as it would be for the situations themselves.

* which criteria *

Experience suggests that we can do quite well, when studying point estimates, with the classical criteria

1) variance (when level does not matter) AND

2) mean square error (when it does).

Certain procedures of improvement, we will learn considerably later, are much more easily applied to (2) than to (1). A convenient way around this difficulty as we will see later, is iterative use of

3) mean square deviation (from some handy base).

Most of what follows would apply to a reasonable variety of other criteria, but, in part to fix our ideas, our discussion will often refer explicitly to one of these criteria.

* the third problem *

So far as we have gone, it seems that the determination of (3) is "close enough" in the case of the Princeton Robustness study (Andrews, et al, 1972) and its continuations (mainly unpublished). This is mainly because of these two facts:

a) the separation of the three situations from one another is quite large,

b) the use of an efficient Monte Carlo allows 640 or 1000 inner configurations to be quite representative of a situation.

* the other problems *

While the selection of (2) is only partially formal, the preselection of a small group of highly polyefficient estimates is both easy and a major step toward the final selection of one -- or perhaps a few -- favored estimates. In this step, it is useful to think in terms of simultaneous "minimization" with respect to several criteria. (See below.)

The problem that has not been previously discussed in an organized way is the bouquet expansion of (1). It is to this question that much of the present account is directed.

4. <u>Inner configurations</u>. The present techniques of Monte Carlo, as applied to finding robust/resistant estimates are most conveniently described in terms of a three-step hierarchy:

1) data set

2) inner configuration, made up of data sets

3) configuration (outer configuration), made up of inner configurations

in which we can expect to know, in using an estimate, what configuration our observed values belong to (as a data set).

In the Monte Carlo, each data set belongs to an inner configuration that is determined by how that data set was generated as part of the raw material for the Monte Carlo. Ordinarily the same set of generated values could, if generated differently, belong to different inner configurations -- though only to one configuration. Thus the raw material for studying estimates in a given finite approximation consists of a list, each element of which combines

1) a data set,

2) information about the inner configuration to which that data set belongs (for the purpose of this specific implementation of the Monte Carlo)

and of instructions of how to use the latter information. In assessing the behavior of an estimate, we use each element of such a list to determine the performance of an estimate, first for the given data set and then for the whole inner configuration which it represents.

* example of inner configurations *

An example may help to fix the ideas, suppose $z_1, z_2, \ldots z_n$ are a sample from the unit Gaussian, Gau(0,1), that $x_1, x_2, \ldots, x_n$ are an independent sample from the unit rectangular Rect[0,1], and that

$\{y_i\}$ where $y_i = \frac{z_i}{x_i}$

is the resulting data set with which we wish to be concerned.

Suppose also that for our purposes the appropriate configuration is the location-and-scale configuration (the lasc configuration) in which all data sets

$\{A + By_i\}$

for a given $\{y_i\}$ make up the configuration.

One natural definition of inner configuration is that it is determined by

1) the lasc configuration of $\{z_i\}$, AND

2) the values of the $\{x_i\}$.

Now, with the $\{x_i\}$ fixed, the $\{y_i\}$ have a joint Gaussian distribution with

$$\text{var}\{y_i\} = 1/x_i^2$$

$$\text{cov}\{y_i, y_j\} = 0$$

$$\text{ave}\{y_i\} = 0$$

so that the (conditional) probability of $\{y_i\}$ in the inner configuration is proportional to

$$\exp\left(-\Sigma x_i^2 y_i^2/2\right)$$

which for

$$y_i = A + By_i^*$$

gives

$$\exp\left(-A^2\Sigma x_i^2/2\right) \exp\left(-AB\Sigma x_i^2 y_i^*\right) \exp\left(-B^2\Sigma x_i^2 y_i^*/2\right)$$

which depends on A, B, and three functions of the data set $\{y_i\}$ and the $\{x_i\}$ used in generating it namely,

$$\Sigma x_i^2$$

$$\Sigma x_i^2 y_i^*$$

$$\Sigma x_i^2 y_i^{*2}$$

In the inner configuration, these three values describe the Gaussian joint distribution of A and B .

Thus we can, if we wish, take either

1) the $\{x_i\}$ and the (lasc configuration of the) $\{y_i^*\}$, OR

2) these three sums and the (lasc configuration of the) $\{y_i^*\}$

as defining the inner configuration. (Choice (2) leads to a much larger inner configuration, larger by $n-3$ parameters.)

The minimum raw material, in this example, for assessing estimate performance, is a list, each of whose elements consists of the n y^*s and the 3 sums PLUS information as to how the sums describe the conditional distribution of A and B in the inner configuration.

II. MULTIPLE MINIMIZATION

We need to think for a little about what is reasonably meant by "minimization" according to two or more criteria. Suppose that for three differently extreme situations, say

G) the pure Gaussian situation,

S) slash, the situation corresponding to the distribution of ratios of independent observations from a centered Gaussian (numerator) and a rectangular distribution abutting zero (denominator),

Ø) one wild, the situation involving $n-1$ observations from $Gau(\mu,\sigma^2)$ and 1 from $Gau(\mu,100\sigma^2)$,

we have the values of

$$V_G(j) = var_G \{candidate\ j\}$$

$$V_S(j) = var_S \{candidate\ j\}$$

$$V_Ø(j) = var_Ø \{candidate\ j\}$$

What are some sensible answers to: "minimize simultaneously with respect to these criteria"?

We shall discuss here, briefly, three approaches that seem reasonable in the context of seeking high quality estimates. (Other approaches may be appropriate for other "simultaneous minimization" situations.)

5. Sayeb minimization. The use of "prices" to combine different criteria (in our case usually one criterion per situation) can have either of two purposes:

1) to simplify the joint minimization, replacing it by a single (combined) criterion (or by a few combined criteria to be considered separately), OR

2) to separate those estimates that "might be preferred" (that are preferred for some set of non-negative prices) from those that "could never be preferred".

The first use requires considerable faith and trust in a single set of prices, but takes us all the way. The second, an "admissability" type of procedure, requires no trust in any particular set of prices, but may leave us with very many estimates that "might be preferred".

Suppose we fix prices, P_G, P_S, and $P_\emptyset$ for the three criteria. We can then consider

$$P_G V_G(j) + P_S V_S(j) + P_\emptyset V_\emptyset(j)$$

as a weighted combination of the three criteria, and ask which j minimizes the result. For any fixed set of prices this reduces the simultaneous minimization to a simple problem.

We need only to solve this problem for enough sets of prices, listing the solutions, to reduce the bouquet of estimates that we need consider, perhaps substantially.

Geometrically, if we consider the convex set generated by the triples

$$\{V_G(j), V_S(j), V_\emptyset(j)\}$$

the list we obtain will consist of those vertices of this convex set at which the perpendicular to at least one tangent plane has all non-negative direction cosines. These vertices define what might be called the <u>lower boundary</u> of the convex set.

* names and concepts *

In the applications to be discussed here the usual pattern is

one criterion for one situation

a fact that may tempt those infatuated with the Bayes model to assert that by "prices" the writer means "a priori probabilities".

As the best witness to what I mean, I assert that this is surely and irretrievably not so.

Those problems where on situation offers two or more criteria, perhaps

1) variance, AND

2) mean square error

show how hard it can be to stretch the Bayes wording to some cases easily treated in terms of prices.

We refer to this process as (complete) Sayeb minimization and the resulting estimates as Sayeb estimates because

1) they are analogous to Bayes minimization and Bayes estimates, AND

2) we most emphatically do NOT want to regard the "prices" as "probabilities".

Complete Sayeb minimization effects a reduction in the size of the bouquet to be considered, but (a) requires considerable calculation and (b) is not likely to reduce the bouquet far enough. (It will always include, for example, the three estimates that perform best in each of three situations separately.)

6. Equal %'s as equally important. One position that is sometimes plausible is to say that an equal % change in behavior under two situations is of equal importance. This choice would make, for instance, the following triples equally desirable:

Raw triple	Triple in %
(2, 20, 200)	(100%, 100%, 100%)
(1, 30, 200)	(50%, 150%, 100%)
(1, 20, 300)	(50%, 100%, 150%)
(2, 10, 300)	(100%, 50%, 150%)
(3, 10, 300)	(150%, 50%, 150%)
(2, 30, 100)	(100%, 150%, 50%)
(3, 20, 100)	(150%, 100%, 150%)

When we come to make this rule of combination properly specific (allowing for the effect on %'s of changing base) we find it reduces, IF we do not wish to specify from what base the %'s are to be calculated, to

1) working with the logs of the criteria, AND

2) using equal Sayeb prices for these logarithms.

We can easily do this (if the change from criterion to log criterion does not affect what we wish to do), the question is whether we wish to do so.

Alternatively, if we are willing to specify the bases B_G, B_S, $B_\emptyset$ from which our %'s are to calculated, then the total percent increase over base is

$$100\left[\frac{V_G(j)}{B_G} + \frac{V_S(j)}{B_S} + \frac{V_\emptyset(j)}{B_\emptyset} - 3\right]$$

minimizing this is equivalent to Sayeb minimization with prices inversely proportional to the bases.

The most natural choice for bases seems to be the best available performance for the corresponding situation

$$B_G = \min_{(\text{bouquet})} \left\{V_G(j)\right\}$$

$$B_S = \min_{(\text{bouquet})} \left\{V_S(j)\right\}$$

$$B_\emptyset = \min_{(\text{bouquet})} \left\{V_\emptyset(j)\right\}$$

The result is a minimization that we should always consider, since it is simple, and not unreasonable. It is, of course, a minimization of average reciprocal efficiency, as is easy to show.

We may not be satisfied with this minimization, since it prefers, for example, (101%, 101%, 120%) to (108%, 108%, 108%), a choice we may or may not think desirable.

A possible variant would be to iterate this minimization with the B's being the variances corresponding to the previous iteration's minimizer. This comes closer to the "minimize sum of logs" approach.

7. Worst case in %. For those who believe they may prefer the (108%, 108%, 108%) choice in the example above, a natural choice is to minimize, instead,

$$\max\left\{\frac{V_G(j)}{B_G}, \frac{V_S(j)}{B_S}, \frac{V_\emptyset(j)}{B_\emptyset}\right\}$$

where, as above, the B's are the best (lowest) values of the criteria when considered one at a time. This is, again, a simple minimization.

Since $V_X(j)/B_X$ is the reciprocal of the efficiency in situation X, this is equivalent to maximizing

$$\min\left\{\text{efficiency}_G, \text{efficiency}_S, \text{efficiency}_\emptyset\right\} = \text{a polyefficiency}$$

a form that is often either convenient or perspicuous, or both.

8. Suggested policy. For the present, the least we ought to seek would seem to be

1) the minimum of "worst case in %" and a list of estimates (with their "worst case" behavior coming close to this minimum, AND

2) the minimum of "total % above base" for the same bases and a list of estimates (with their "mean % up" behavior) coming close to this minimum.

Such a policy, with our present degree of insight, seems to offer a good basis from which to stir in other considerations, like ease of use and ease of description. (We might, for example, like to do the formal analysis just described for each of several bouquets successively restricted to greater and greater ease.)

III. IMPROVING THE BOUQUET: LINCOMS

In discussing how to improve the bouquet, we will discuss three specific approaches to improvement by extension:

1) forming convex lincoms (forming convex linear combinations of present candidates),

2) compartmenting for level adjustment, in which we divide data sets (really configurations) into compartments according to observable characteristics and then make adjustments to an estimate's value according to the compartment in which it falls,

3) compartmenting for selection, in which we divide configurations (data sets) into compartments, and then use a composite estimate defined by using a (possibly different) candidate estimate in each compartment.

All of these approaches have shown promise.

The general pattern we shall follow in discussing these three approaches will be:

a) What we do.

b) How much good it seems to do.

c) How we start.

d) Where we might go next.

9. Forming linear combinations. If estimate i_1 and estimate i_0 are in our bouquet, they must already be described. Thus, for any given θ, it is easy to describe

$$\text{estimate } i_\theta = (\theta)(\text{estimate } i_1) + (1-\theta)(\text{estimate } i_0)$$

whose variances, say for G and S, are

$$V_G(i_\theta) = \theta^2 V_G(i_1) + 2\theta(1-\theta)\, C_G(i_1, i_0) + (1-\theta)^2 V_G(i_0)$$

and

$$V_S(i_\theta) = \theta^2 V_S(i_1) + 2\theta(1-\theta)\, C_S(i_1, i_0) + (1-\theta)^2 V_S(i_0)$$

where $C_G(i_1, i_0)$ and $C_S(i_1, i_0)$ are the covariances in the two situations. Here we can take any θ with $-\infty < \theta < +\infty$, thus obtaining a whole pencil of estimates.

* along one pencil *

Let θ_G be the value of θ that minimizes the G-variance. Since the G-variance of i_θ is quadratic in θ, we have

$$V_G(i_\theta) = a_G + b_G^2(\theta-\theta_G)^2$$

and similarly

$$V_S(i_\theta) = a_S + b_S^2(\theta-\theta_G)^2$$

so that

$$\sqrt{V_G(i_\theta)-a_G} = \pm b_G(\theta-\theta_G) = b_g|\theta-\theta_G|$$

$$\sqrt{V_S(i_\theta)-a_S} = \pm b_S(\theta-\theta_S) = b_s|\theta-\theta_S|$$

so that the path of our pencil in the root-excess-variance plane -- the plane with coordinates

$$\left(\sqrt{V_G(i_\theta)-a_G}\ ,\ \sqrt{V_S(i_\theta)-a_S}\right)$$

is made up of three line segments that obey Snell's law when touching a coordinate axis. Thus the outgoing ray $(\theta \to +\infty)$ is parallel to the incoming ray $(\theta \to -\infty)$.

In the $(V_G(i_\theta), V_S(i_\theta))$ plane, the pencil will, as is easily seen, be a convex curve; our interest in it is thus confined to the arc between the points $\theta=\theta_G$ and $\theta=\theta_S$.

If we consider three (or more) situations, we get, for example

$$\sqrt{V_G(i_\theta)-a_G} = b_g|\theta-\theta_G|$$

$$\sqrt{V_S(i_\theta)-a_S} = b_s|\theta-\theta_S|$$

$$\sqrt{V_\emptyset(i_\theta)-a_\theta} = b_\emptyset|\theta-\theta_\emptyset|$$

and in root-excess-variance 3-space we get 4 (in general k+1) line segments where now <u>one</u> direction cosine is multiplied by -1 at each touching of a coordinate hyperplane. Again the outgoing ray is parallel to the incoming one.

In the

$$(V_G(i_\theta), V_S(i_\theta), V_\emptyset(i_\theta))$$

3-space we get a curve, where our interest is confined to the smallest interval containing all 3 (in general all k+1) of θ_G, θ_S, $\theta_\emptyset$.

* joint minimization *

If we are doing Sayeb minimization, with prices P_G, P_S and $P_\emptyset$ we wish to minimize

$$P_G(a_G+b_G{}^2(\theta-\theta_G)^2) + P_S(a_S+b_s{}^2(\theta-\theta_S)^2) + P_\emptyset(a_\emptyset+b_\emptyset{}^2(\theta-\theta_\emptyset)^2)$$

which occurs when (differentiating and canceling 2's)

$$P_Gb_G{}^2(\theta-\theta_G) + P_Sb_S{}^2(\theta-\theta_S) + P_\emptyset b_\emptyset{}^2(\theta-\theta_\emptyset) = 0$$

when, that is

$$\theta = \frac{P_Gb_G{}^2\theta_G + P_Sb_S{}^2\theta_S + P_\emptyset b_\emptyset{}^2\theta_\emptyset}{P_Gb_G{}^2 + P_sb_S{}^2 + P_\emptyset b_\emptyset{}^2} = \theta_P$$

Thus, if we know the corresponding covariances as well the variances, we can find not only the θ_G , θ_S and $\theta_\emptyset$ but the b_G^2 , b_S^2 and $b_\emptyset^2$. And, if we also know the prices P_G , P_S and $P_\emptyset$ we can find θ_P .

To do "worst case in %" minimization along such a pencil requires only slightly more calculation, at least for 3 situations, since there are only two ways in which the worst % of 3 can be minimized:

a) one % is at minimum, and the other 2 are smaller (we must look at θ_G, θ_S or $\theta_\emptyset$).

b) two of the % are equal and the third is at least as small (requires three solutions of simple quadratic equations for the points of equality, each easily tested).

10. Experience with convex lincoms. Experience is mainly confined to looking at only $\theta = 1/2$, rather than at jointly minimizing combinations, although the latter does not seem difficult. One reason for going slowly is that if, for example, we start with a bouquet of 85 estimates, we obtain

$$\binom{85}{2} = 3570$$

50-50 lincoms. This is already large enough to require computer handling.

So far, the study of convex lincoms has been of greatest help in the early-intermediate stages of bouquet improvement. Here it has been most useful as a source of encouragement, as a sign that better things CAN be done. One reason for this is simple: A finding that a 50-50 combination of two estimates, naturally described in qualitatively different ways, works well is not easy to convert into a suggestion of new candidate estimates that are reasonably simply described.

11. Starting on convex lincoms. So far as experience goes, we suggest starting with a reasonably large bouquet of quite differently described but rather highly polyefficient estimates.

12. Where we might go next. The only two natural steps forward would seem to be

a) trying the jointly minimizing combinations instead of just the 50-50 ones

b) trying only the convex lincoms of one (or a few) highly favored estimates with a larger, more diverse, body of other estimates.

IV. IMPROVING THE BOUQUET: LEVEL ADJUSTMENT

13. Compartmentation. We need to fix in our minds that, in compartmentation, in which we divide all data sets into compartments in terms of recognizable characteristics, inner configurations, and even (outer) configurations, remain together, each going as a whole into a particular compartment. We are now dividing on the basis of only what we can see, and not at all on how a data set happened to be generated in preparing for a Monte Carlo.

It is probably reasonable to think of 5 to 10 compartments as usual, though any examples we give here are likely to be for 2 (or 3) compartments as a matter of simplicity.

One way to think about compartmentation, but NOT the only way, is to think of it as a way to partially sort out the situations involved. We might, for instance, use some measure of apparent tail-stretchedness to make one squeezed-tail and one stretched-tail compartment. If we generate data sets as Gaussian samples, we would expect most to fall in the squeezed-tail compartment. If we generate data sets as samples from a slash distribution, we would expect most to fall in the stretched-tail compartment. Thus it may be quite reasonable to allow its compartment to influence what happens to a data set.

The same might be true, of course, if we considered a compartmentation where equal numbers of both Gaussian and slash samples fall into each compartment. Some characteristic of our data sets (here, samples) which is unrelated to whether their source was Gaussian or slash, might be important. We can take sorting out sources as a leading case but we dare not, so far as we now know, take it as the only important case.

14. Compartmentation for level adjustment. Here we work with a single estimate. We suppose that the following fractions of all data sets from sources A, B and C,

q_A, q_B, and q_C fall in compartment 1

r_A, r_B, and r_C fall in compartment 2

with average values for our estimate, measured as deviations from the "correct" value, of

x_A, x_B, and x_C in compartment 1,

y_A, y_B, and y_C in compartment 2,

and variances, pooled within compartment, of

$$V_{in}(A)\ ,\ V_{in}(B)\ ,\ V_{in}(C)\quad .$$

* the mean square error *

The mean square errors for the original estimate will be

$$MSE_A = V_{in}(A) + q_A x_A^{\ 2} + r_A y_A^{\ 2}$$

$$MSE_B = V_{in}(B) + q_B x_B^{\ 2} + r_B y_B^{\ 2}$$

$$MSE_C = V_{in}(C) + q_C x_C^{\ 2} + r_C y_C^{\ 2}$$

If now we make level adjustments of $-\Delta$ and $-\delta$ in the two compartments, so that $x_A \to x_A - \Delta$, $x_B \to x_B - \Delta$, ..., $y_C \to y_C - \delta$, we will have, for

$$(P_A)(MSE_A) + (P_B)(MSE_B) + (P_C)(MSE_C)$$

a quadratic in Δ and δ whose minimum is attained when

$$\Delta = \frac{P_A q_A x_A + P_B q_B x_B + P_C q_C x_C}{P_A q_A + P_B q_B + P_C q_C}$$

$$\delta = \frac{P_A r_A y_A + P_B r_B y_B + P_C r_C y_C}{P_A r_A + P_B r_B + P_C r_C}$$

* minimizing variance *

We need now to notice that minimizing mean square errors (a) is not the same as minimizing variances, but (b) can be used as a step in reducing variances.

For a simple numerical example of the difference, take three compartments and two situations, A and B, where the three fractions are q, r, and s, and the average values x, y and z with:

Source A (situation A): $q_A = .05$, $r_A = .20$, $s_A = .75$,
$x_A = 0$, $y_A = 1$, $z_A = 3-u$

Source B (situation B): $q_B = .75$, $r_B = .20$, $s_B = .05$,
$x_B = 0$, $y_B = 0$, $z_B = 0-u$

where u , which we are going to vary, is the additive adjustment to the values in compartment C .

We clearly have

$$MSE_A = V_{in}(A) + .20 + .75(3-u)^2$$

$$MSE_B = V_{in}(B) + .05(-u)^2$$

which are minimized, respectively, at $u_A = 3$ and $u_B = 0$, and average values

$$.20 + .75(3-u)$$
$$.05(-u)$$

whence

$$V_A = V_{in}(A) + .20 + .75(3-u)^2 - (.20 + .75(3-u))^2$$

$$= .95 - .825u + .1875u^2$$

$$V_B = V_{in}(B) + .05(-u)^2 - \big(.05(-u)\big)^2 = V_{in}(B) + .0475u^2$$

which are minimized respectively, at $u = 2.2$ and $u = 0$.

Thus, if we are minimizing variance, we have no interest in u's outside [0, 2.2]. The most plausible values of u for minimizing MSE's are NOT in this range, as the following short table suggests:

u	MSE_A	MSE_B
2.0	$V_{in}(A) + .95$	$V_{in}(B) + .20$
2.2	$V_{in}(A) + .68$	$V_{in}(B) + .24$
2.5	$V_{in}(A) + .39$	$V_{in}(B) + .32$
3.0	$V_{in}(A) + .20$	$V_{in}(B) + .45$

Unless $V_{in}(B)$ is considerably smaller than $V_{in}(A)$ we are likely to prefer a u somewhere near 2.5 (for minimizing MSE's) which is wastefully large for minimizing variances.

* iterating to variance minimization *

If our real concern is with variance rather than with mean square error -- if we do not care about the average value of our estimate (this really happens sometimes) -- we can proceed iteratively if we wish. Having minimized the mean square error for some target value, we can adopt the resulting average values, one for each source, as our handy new target value, and minimize mean square deviations from these values, again, and again, and again... . The process must converge since the sum of the mean square deviations is ever decreasing. At convergence we are taking the mean square deviations about the average values, so that we are indeed dealing with the variances.

15. Experience with level adjustment. Experience is not extensive, but indicates that such processes can, and at times do, provide appreciable improvements in the later stages of bouquet enrichment. This seems to be particularly so when

1) the compartments are defined in terms of some quantitative slicing characteristic, AND

2) the resulting compartment adjustments are used to suggest a continuous function of the slicing characteristic that can be used to provide a smooth level adjustment.

These techniques have been used in the past year both to improve the available performance of free widthers for individual situations and to improve the polyefficiency of a free widther as calculated for the usual (G, S, and Ø) three situations (Schwartzschild, 1978).

16. Where to start. For the present, if we are working with the three by-now-conventional situations, it seems natural to start with quantitative characteristics that tend to separate the (neutral-tail) Gaussian source from the (stretched-tail) slash and one-wild situations. Here the results of Rattner (1977) indicated that

javaj ln s^2

where "javaj" stands for "jackknife estimate of variance for the jackknifed value of" and s^2 is the usual fraction of the sum of squares of deviations, might be a good basis for defining compartments. (This is what was used for the free-widther improvements just described.)

We need more experience with alternative compartmentations in diverse circumstances before we can be very explicit. Clearly the general rule is to try to identify what really makes a difference, and then isolate it about as well as an observable characteristic can do this.

17. Where to go next. If we consider our original estimate and the modified estimates

$$\text{estimate}_1 = \text{original estimate PLUS} \begin{cases} 1, & \text{in compartment 1} \\ 0, & \text{elsewhere} \end{cases}$$

$$\text{estimate}_0 = \text{original estimate PLUS} \begin{cases} 1, & \text{in compartment 2} \\ 0, & \text{elsewhere} \end{cases}$$

we can see that "level adjustment" can be equally well described as "using, in compartment j, a convex linear combination of estimate_j and the original estimate". Clearly we could consider replacing estimate_j by any other interesting estimate.

While it is not clear how profitable such procedures would be, it would be interesting to find out.

V. IMPROVING THE BOUQUET: SELECTION

18. Compartmentation for selection. Suppose have m clearly defined estimates. Any rule of the form: "First determine in which compartment the data set falls, and then apply the estimate specified for that compartment." is now a well defined estimate. There are

$$m^k$$

such selections, m^k rules assigning one of the m estimators to each of the k compartments.

We could, in principle, put each of these m^k estimates into an enlarged bouquet -- there are only a finite number of them. For what seems like a realistic pair of numbers,

20 estimators and 7 compartments, however, there would then be 1.3×10^9 selections. Dealing with these one at a time is not likely to be very useful.

We need to restrict the problem to the point where something can be done with it.

It seems likely that we will be wise to begin by focusing on criteria that are additive over compartments, for which, if j is a general compartment and B is any situation (ideal or finitely approximate), the identity

$$\text{Criterion}_B = \sum_j (\text{criterion}_B \text{ at } j)$$

holds for any candidate estimate.

Such criteria seem to usually, if not always, to be ones which are defined by a sum (finite actuality) or integral (ideal theoricity) over "all data sets".

Examples of such criteria are the MSE (mean square error) which is proportional to

$$\sum (\text{deviations from truth})^2$$

and the MSD (mean square deviation) which is proportional to

$$\sum (\text{deviations from handy})^2$$

where "handy" is at our choice. (The last form clearly suggests interative calculation where "handy" depends on the result of the last iteration. We have described how such an iteration can be used to minimize variance.)

19. Selection for multiple criteria. So far we have only mentioned a single criterion explicitly. What effect does the number of criteria have?

If we have only a single criterion, a single situation, picking a selection is simple -- we have only to pick the best estimate in each compartment. The same is true if we wish to do Sayeb minimization with a fixed set of prices, no matter how many situations are involved.

If there are just two situations, we can afford a complete Sayeb minimization. Only the ratio of the two prices matters, and we can do fixed-price minimization for a sufficiently closely spaced values of the ratio. (We need at most $km+1$ values if we already know where to choose them.) We can then apply any combination of the two criteria to the resulting list.

If we have many criteria, complete Sayeb minimization seems quite out of the question. There can be very many vertices on the lower boundary of the convex region defined by the m^k selections. All is not lost, however, since at least two approaches remain feasible:

1) we may take prices inversely proportional to the values found in the previous iteration, thus approaching the minimum-sum-of-% solution.

2) we may take advantage of a result of Harold Kuhn's (1977) which reduces the minimization of

$$\max\left\{\text{various additive criteria}\right\}$$

to a linear programming problem.

While the LP problem is not trivial, its size is only of order mk and is quite within range of the available techniques. Thus, for additive criteria, we can anticipate doing selection for either of what we regard as the principal types of combined criteria.

What we are doing here has some similarity to some of Hogg's interesting proposals (e.g. Hogg, Uthoff, Randles and Davenport 1972, Randles and Hogg 1973) for the centering problem. It is important to understand both the similarities and the differences.

Hogg used a tail-stretch-tail-squeeze sample compartmentor (namely the sample relative fourth moment) and for each compartment suggested using an estimate that would have good performance in sample from a distribution for which the corresponding population compartmentor would be somewhere near the center of the compartment.

Were we to apply the procedure just discussed we would appear with a fairly large bouquet of estimators, possibly including some of those suggested by Hogg, and a moderately large number of situations. We would then try to find out which estimate it was empirically best to use in a given compartment, whether or not its use was suggested by large-sample results or equating sample quantities to population quantities.

The flavor is rather similar, but both purposes and procedure are quite different.

20. Experience with selection. What experience we have here is with informal techniques rather than with the machinery just sketched. In the centering problem, this type of approach, using what we now regarded as unlikely to be an excellent compartmenting variable, increased the tri-efficiency from 92% to 93% (the largest known value before John's increase to 94%+ reported at this workshop). While we now know enough more about widthing to throw the centering problem open to further development, it seems likely that compartmentation for selection

1) will remain an effective way of adding a final polish, AND

2) will often provide the basis for suggesting new or revised forms for estimates.

21. Where to start. It is not clear just what kinds of sets of estimates are most likely to be good starts for compartmentation for selection. In the centering case mentioned above, the estimates considered differed in the value of one continuous parameter. Such sets are certainly very likely to lead to clear suggestions of revised candidates, as are sets where the only differences are the values of a few continuous parameters.

What we will be able to do with quite different kinds of sets is still unclear.

22. Where to go next? We need to gain experience with compartmentation for selection, as presently envisaged, before we try to answer this question.

VI. FINDING CANDIDATES

23. Starting in general. If we are attacking a wholly new problem area, our early tasks are to find:

1) excellent candidates for each single situation (as the basis for calculating apparent efficiency, etc.)

2) good candidates for the combined situations (as the basis for improved bouquets of estimates)

3) good compartmentors (as aids in bouquet improvement).

What can we say about how to do this -- how to begin exploration of a new wilderness?

24. Single situations. The classical approaches to the single situation cases focus around maximum likelihood. While this is far from being a likely route to good compromises, there is increasing evidence (e.g. Hoaglin 1971, Schwartzschild, 1978) that maximum likelihood, particularly when adjusted, does very well in single situations. Thus it would seem natural to begin with either maximum likelihood estimates (likely to require iteration) or score-based analogs (or both), AND to plan to, with the aid of the best compartmentings at hand, to at least use compartmenting for level adjustment to improve these starting estimates.

At the beginning, there are not likely to be enough strong candidates to make either of the other two major improvement processes worthwhile. But we must remember that the whole process is deeply iterative and bootstrap-like. So as we learn more about good compromises, and expand our bouquet of estimates, we should plan to come back, from time to time, to the single situations.

25. Compromises. Both insight and luck help in finding good compromise candidates. What is probably the most useful heuristic involves combining the single excellent estimates already found for the single situations, using the principle:

"When in doubt, give a lower weight"

Here the (differential) weight given to y_i in an estimate can be reasonably taken as

$$(\text{number of observations})\frac{\partial(\text{estimate})}{\partial y_i}$$

The more one learns about a problem, and the more compartmentation (and, perhaps, lincoming) is used to expand and improve the bouquet, the more likely we are to have useful ideas for new kinds of estimates.

26. Compartmentors. In those elementary problems where location and scale are free, and cannot be used for compartmentation, the natural aspects of a data set on which to focus next (we believe without proof) are:

1) skewness

2) elongation (positive elongation = appearance of stretched tails).

We do not know, at this time, how much is to be gained by compartmenting based on apparent skewness. Largely, one might suspect, because of assumptions of symmetry for our situations, we have not tried such possibilities.

There is a moderate amount of experience with elongation compartmentors. In the centering problem, the compartmenting -for-selection work mentioned earlier used the studentized maximum deviate. As noted above, Rattner's work suggested javaj ln s^2, which has already proved profitable in the widthing problem.

In the interval centering problem, work by Karen Kafadar indicates that, for w-estimates, the sum of weights is likely to be a good compartmenting variable.

It seems likely that we have only begun to scratch the surface.

VII. OTHER DIRECTIONS FORWARD

27. Diversified preMonteCarlo's. Our earlier discussion has emphasized the separation of Monte Carlo work into sections:

1) the preparation of basic materials, for example data sets with the numbers needed to make them relevant to inner configurations

2) evaluation of estimate results using these basic materials

3) summarization of estimate behavior

4) incisive data analysis of the estimate results with an aim toward bouquet expansion and improvement.

So long as our attention was mainly devoted to (3), the case for having weighted basic materials, for example, could only rest upon the efficiencies of stratified sampling. But, as our view of Monte Carlo studies shifts from the passive toward the active, our interest in compartments is not at all bounded by contributions to the variances of estimated variances. We may need enough inner-configured data sets to yield understanding and guidance from a compartment-situation problem that arises only rarely.

Both separate files of basic materials for such compartment-situation combinations, and files of weighted basic materials in which low weights in such compartments are reflected by more (weighted and inner-configured) data sets are likely to be useful.

28. Polyregression. If we seek for an at least semi-formalized scheme within which much of what we discussed in part III can be formulated, the natural thing is an analog of regression adjustment, which it is convenient to call polyregression.

In conventional regression adjustment, we have y's , and with each a certain number of x's , whose means are either (a) known exactly, or (b) known to greater accuracy. As usually described, we do a regression of y on the x's , and then think of using the values of

$$y-b_1x_1-b_2x_2, \ldots -b_kx_k$$

to estimate

$$\text{ave}\{y\}-b_1\text{ave}\{x_1\}-b_2\text{ave}\{x_2\}-\ldots-b_k\text{ave}\{x_k\}$$

which estimate we adjust by adding

$$b_1\text{mean}\{x_1\} + b_2\text{mean}\{x_2\}+\ldots+b_k\text{mean}\{x_k\} .$$

An equivalent and often more useful description is that we regress y on the carriers

$$x_1-\text{mean}\{x_1\},\ x_2-\text{mean}\{x_2\},\ \ldots,\ x_k-\text{mean}\{x_k\}$$

and, then since these carriers all have mean zero, use the mean of the regression residuals to estimate the average of y .

In polyregression for estimate improvement, y stands for the value of an estimator in which we are interested, and the x's are functions of configuration (estimators of zero are a special case). We take as our new estimator the residual of y after polyregression on the x's . (Since the change of estimate is a function of configuration alone the changed estimate may be regarded as estimating "the same thing".)

In the calculation of the polyregression itself, we strive to minimize several criteria, either through chosen prices or other combinations. Exhibit 1 shows how all three processes we have considered, and some we haven't, fit into the polyregression structure. Clearly there are interesting possibilities for the future.

Exhibit 1

Instances of polyregression for estimate improvement

Approach	Uses polyregression on
Lincoms	Differences of estimates of the quantity.*
Level adjustment	Indicator functions of compartments.
Selection	Product of difference of estimates and indicator functions (with the polyregression coefficients limited to 0 and 1).
Future 1	1 and compartmentor.**
Future 2	Same within a compartment (zero elsewhere).
Future 3	Polynomials, or more interesting functions of the compartmentor.
Future 4	As for selection, with unrestricted polyregression coefficients.
General 1	Function of configuration
Location 1	Function of configuration times estimate of scale (useful when both location and scale are free).

*In some problems -- e.g. location with known scale, e.g. log scale -- the difference of two estimates of the desired aspect is a function of configuration. In others, the specification includes higher parameters, and we have to use estimates of them appropriately to convert the differences of two estimates into a function of configuration -- e.g. location with scale unknown, where we have to divide by an estimate of scale, (compare the "location 1" line.)

**The "compartmentor" is any continuous characteristic that could have been used to define the compartments we are using.

29. Hybrid estimates. Consider first linear combinations of ordered values $y_1, y_2, \ldots, y_n$ and of their gaps $g_1, g_2, \ldots, g_{n-1}$. A small amount of algebra shows us that

$$w_1y_1+w_2y_2+\ldots+w_ny_n \equiv w_1g_1+(w_2+w_1)g_2+\ldots+(w_{n-1}+\ldots+w_1)g_n$$
$$+(w_n+\ldots+w_1)y_n$$

$$v_1g_1+v_2g_2+\ldots+v_{n-1}g_{n-1} \equiv -v_1y_1+(v_1-v_2)y_2$$
$$+\ldots+(v_{n-2}-v_{n-1})y_{n-1}+v_{n-1}y_n$$

so that, except for the one term $(w_n+\ldots+w_1)y_n$, the expressions

$$\Sigma w_iy_i$$

and

$$\Sigma v_ig_i$$

are equivalent for constant weights. Thus any rank-based linear combination

$$\Sigma w_iy_i$$

can, in particular, be put in an equivalent form

$$v\dot{y} + \Sigma v_ig_i$$

where $\dot{y}$ is the median y .

If we ponder on these relations when the w_i and v_i are no longer constant, we are led to interesting possibilities. Any w-estimate, of the form

$$\frac{\Sigma w_i(y)\cdot y_i}{\Sigma w_i(y)}$$

where

$$w_i(y) \equiv w(y_i)$$

can be put in the form (y stands for all the y_i)

$$\dot{y} + \frac{\Sigma v_i(y)g_i}{\Sigma w_i(y)}$$

where (for n odd -- the situation differs only in detail for n even)

$$v_i(y) = -w(y_1)-w(y_2)-\ldots-w(y_i) \qquad \text{for } i < \frac{n}{2}$$

$$= w(y_n)+w(y_{n-1})+\ldots+w(y_{i+1}) \qquad \text{for } i > \frac{n}{2}$$

This suggests trying both

$v_i(y)$ = a more general function of the y's and i

and

$w_i(y)$ = a more general function of the y's and i

(These only differ, of course, when we at least partially specify the form of these more general functions.)

Thus we may be able to develop estimates that combine the advantages of w-estimates or m-estimates on the one hand and the (lesser) advantages of rank estimates on the other.

REFERENCES

1. D. F. Andrews, P. J. Bickel, F. R. Hampel, P. J. Huber, W. H. Rogers, and J. W. Tukey (1972), Robust Estimates of Location, Princeton University Press.

2. D. C. Hoaglin (1971), Optimal invariant estimation of location for three distributions and the invariant efficiencies of other estimates. Ph.D. Thesis, Department of Statistics, Princeton University.

3. R. V. Hogg, V. A. Uthoff, R. H. Randles, and A. S. Davenport (1972), On the selection of the underlying distribution and adaptive estimate, JASA, 67, pp. 597-600.

4. R. H. Randles, and R. V. Hogg (1973), Adaptive distribution-free tests, Commun. Stat. 2, pp. 337-356.

5. Z. Rattner (1977), A Monte Carlo study: jacknife width estimation of robust statistics. Unpublished Senior Thesis, Department of Statistics, Princeton University.

6. M. Schwarzschild (1978), New observation-outlier-resistant methods of spectrum estimation, Ph.D. Thesis, Department of Statistics, Princeton University.

The author was supported by a contract with U. S. Army Research Office (Durham) No. DAAG29-76-G-0298.

Department of Statistics
Princeton University
Princeton, New Jersey 08540

Robust Techniques for the User

John W. Tukey

1. Introduction. This is a "pitch" for users. It has been worded and formatted with that in mind. The original form was prepared during the night of 11-12 April 1978, in the fear that the first day's papers at the workshop were likely to miss their targets -- users. Since the topics on April 12 were overtly more sophisticated, there has been no attempt to expand the coverage further.

2. The overall message. THE CLASSICAL METHODS -- means, variances and least squares -- ARE UNSAFE.

Sometimes they are good, sometimes they are not.

3. A comment. Just which rob/res (robust and resistant) methods you use is NOT important -- what IS important is that you use SOME.

It is perfectly proper to use both classical and robust/resistant methods routinely, and only worry when they differ enough to matter. BUT when they differ, you should think HARD.

4. What to seek. TRYING to keep things looking "LIKE" classical procedures leads to CONFUSION.

What you need is a reasonably self-consistent set of procedures that are reasonably easy to use and reasonably easy to describe. For one such, see Chapter 10 of Mosteller and Tukey, 1977.

ISBN 0-12-438150-2

5. What were the high points on April 11? Hogg told you to ALWAYS run BOTH rob/res and classical procedures, (note which one I put first.)

Andrews told you that classical residuals COVER UP like WATERGATE. To learn what is going on often requires looking at ROB/RES residuals.

Agee told you that a large highly-stressed computing shop FINDS rob/res techniques very WORTH WHILE. Those who may have to get a million values "today" and give the answer "tomorrow" cannot wait to look for trouble; they have to have trouble-dodging procedures built in. (I spent a week during World War II cleaning up test data on a device known only to me as "Mickey" -- it turned out to be the first land-based AA radar. A week was too long.)

Johns told you that new ideas can still push the quality of rob/res estimates upward by unexpected amounts. This means that the 1981 recommended versions will be different from, and somewhat better than, the 1977 recommended versions. But what matters to the USER is that either is FAR SAFER than the 1813± version -- LEAST SQUARES.

6. Three things you need to be clear about:

(a) efficiency. EFFICIENCY for the USER needs to be interpreted quite differently for the user than for the tool forger. (We need both interpretations.) ALL efficiencies between 90% and 100% are NEARLY the SAME for the USER.

Alternate feeding of bodies of data to two statisticians, one of whom uses a 90% efficient estimate, the other using a 100% efficient estimate, followed by comparing each's estimates with the corresponding truths, has to involve like 3000 bodies of data before we can prove which is which. Nothing methodological that takes this much data to check is likely to be important.

The TOOL-FORGER, on the other hand, SHOULD pay attention to another 1/2% of efficiency:

1) because step-by-step growth can take us a long way,

2) because learning how to do even a little better "here" may lead to widening the area where we can do well enough,

3) because looking for "a little more" sometimes leads to an EASIER alternative.

7. Three things you need to be clear about: (b) order statistics and their linear combinations. Whatever impressions others give;

1) order statistics and their linear combinations are NOT the whole story.

2) sticking to them exclusively is like sticking to EPICYCLES after Copernicus.

3) order statistics and their linear combinations are NOT the easiest rob/res statistics to use.

8. Three things you need to be clear about: (c) multiple challenges. The typical "proof of the pudding is in the eating" test subjects statistical techniques to diverse challenges and assesses how well they do. Three styles are important:

1) OLD STYLE. One challenge, algebraic assessment of efficiency (e.g. least squares).

2) MIDDLE STYLE. Lots and lots of challenges, algebraic-analytic assessment -- of validity (conservatism) alone (e.g. non-parametric techniques).

3) NEW STYLE. A few carefully selected (to be overextreme) challenges, computer simulation assessment of (validity and) efficiency (e.g. today's rob/res methods).

In the so-called Princeton Robustness Study (Andrews, et al 1972) we used twenty-odd different challenges; three turned out to be CRUCIAL. (If a boxer can stand up to hard blows:

1) on the jaw,

2) at the solar plexus,

3) on the back of the neck ("rabbit punch")

he is likely to stand up to almost anything.)

Frank Hampel thinks we should consider 5 or 6 such overextreme challenges instead of only 3. He may well be right -- but if he is and if we change, the answers will not change violently.

9. Query and answer. I was asked yesterday "WHAT RIGHT" we had to TINKER with the TAILS of the sample.

The answer is simple: WE HAVE PUT MANY ESTIMATES TO THE TEST OF VERY DIVERSE CHALLENGES AND THE ONE WE NOW RECOMMEND performed well AGAINST all OF THESE CHALLENGES.

The fact that we have heuristics -- some of them called asymptotic theorems -- that DO explain the type of thing that PROVES to WORK is SECONDARY for the USER (though closer to primary for the TOOL-FORGER).

10. Good tools are not naive. Most of us would, I suspect, be prepared to believe that the design choices in a klystron or a large aircraft were unlikely to be obvious. Fewer of us would recognize that the choices of an ax or a saw are also not obvious. (If you had never seen an ax, how would you have shaped its handle?)

Today some of our best statistical techniques have reached the design sophistication of an ax or saw:

1) you have to try them to be sure of their value

2) after you know what works, you can convince yourself, with enough effort, it was sensible after all.

Tomorrow we may expect still better procedures for which (1) holds but NOT (2). (When they put the obviously aerodynamically bad blisters on the World War II Catalina (PBY) flying boat, the top speed went UP! Easily observed; hard to explain. This is what you should be getting ready for!)

REFERENCES

1. D. F. Andrews, P. J. Bickel, F. R. Hampel, P. J. Huber, W. H. Rogers, and J. W. Tukey (1972), Robust Estimates of Location, Princeton University Press.

2. F. Mosteller, and J. W. Tukey (1977), Data Analysis and Regression: A Second Course in Statistics, Addison-Wesley Publishing Company, Reading, MA. Especially Chapter 10.

The author was supported by a contract with U.S.Army Research Office (Durham) No. DAAG29-76-G-0298, Department of Statistics Princeton University, Princeton, N. J. 08540

Application of Robust Regression to Trajectory Data Reduction

William S. Agee and Robert H. Turner

INTRODUCTION

Robust Statistics provides a new approach to data editing in trajectory data reduction. Data editing, whose function is to deal with wild observations, has been a most frustrating problem for the data analyst. The use of robust statistics has been highly successful, much more so than previous methods, in dealing with this problem. There are several applications of robust statistics to data editing in trajectory data reduction. The applications considered here are:

Data Preprocessing
Instrument Calibration
N-station Cine Solution
N-station Radar Solution
Filtering

Before describing these applications we need to answer: What are robust statistics and how do robust statistics apply to data editing? In answer to the first part of the question robust statistical methods are those which tend to retain their desirable properties under at least mild violations of the assumptions under which they were derived. Possibly a more useful description of a robust statistical procedure is one which will perform well under a variety of underlying distribution functions or in the presence of observations from contaminating distributions. In answer to the second part of the question we are probably not very concerned about the performance of data reduction procedures under a wide variety of underlying distribution functions of the observations but are mainly concerned about the performance of our methods in the presence of observations from contaminating distributions, i.e., outliers. Thus, in data

ISBN 0-12-438150-2

reduction we are interested in the development of robust statistical methods which are highly outlier resistant. In data reduction we are usually interested in estimating the parameters in some postulated linear or nonlinear model of a process. Thus, in data reduction we are specifically interested in developing methods for linear and nonlinear regression which are insensitive to a large percentage of outlying observations. Many sources of outliers are present in trajectory measuring systems. Without going into any detail, these sources may be broadly grouped into the categories of equipment malfunction, outside interference, and human error.

The usual methods of least squares, optimally weighted least squares, maximum likelihood, etc., used in data reduction for estimating parameters in a regression model provide useless estimates in the presence of outliers. To quote Huber [1], "even a single grossly outlying observation may spoil the least squares estimate and moreover outliers are much harder to spot in the regression case than in the simple location case."

Although the history of robust estimation goes back to the 19th century, the development of robust regression methods is just currently becoming a popular topic for statistical research. Some of the earliest methods for robust regression were developed in the 1950's, notably the methods reported by Brown and Mood [9] and by Theil [5]. Robust estimation methods have been classified by Huber [1] and [2]. Huber's classifications are termed L-estimates, M-estimates, and R-estimates. The L-estimates are estimates which are linear combinations of the order statistics. The R-estimates are estimates derived on the basis of rank tests. Of the three classifications for robust estimates given by Huber we shall mainly be concerned with M-estimates for application to data reduction.

Given the linear model

$$y_i = \sum_{j=1}^{P} x_{ij}\theta_j + e_i \ , \qquad i=1,N \ ; \tag{1}$$

the regression parameters θ_i are to be estimated. The M-estimates of θ_i minimize

$$\sum_{i=1}^{N} \rho(y_i - \sum_j x_{ij}\theta_j) \ , \tag{2}$$

where $\rho(\cdot)$ is some function which is often convex. Differentiating with respect to θ leads to

$$\sum_{i=1}^{N} X_i^T \psi(y_i - \sum_j x_{ij}\theta_j) = 0 \ , \tag{3}$$

where

$X_i^T = \mathrm{col}(x_{i1}, x_{i2}, \text{- - -}\ x_{ip})$

and $\psi(\cdot)$ is the derivative of $\rho(\cdot)$. (3) is the analog of the normal equations in least squares regression. The estimate which results from solving (3) is called an M-estimate. Rather than specifying the function ρ, M-estimates are usually described by specifying the function ψ. If $f(y;\theta)$ is the probability density function underlying the observations, and if

$$\psi = \frac{\partial f(y;\theta)}{\partial\theta} \Big/ f(y;\theta) ,$$

then the M-estimate obtained is the maximum likelihood estimate. Since the function ρ is usually not homogeneous, as it would be in least squares the M-estimates obtained would usually not be scale invariant. Hence, to force scale invariance we minimize

$$\sum_{i=1}^{N} \rho\{(y_i - \sum_j x_{ij}\theta_j)/s\}, \tag{4}$$

where s is some measure of dispersion of the residuals, $y_i - \sum_j x_{ij}\theta_j$. The measure s also needs to be a robust measure of dispersion.

Several ψ functions have been proposed in the literature. Basically those ψ functions fall into two classes, the redescending type and non-redescending type. We will only consider one member of each of these classes in this report. The original ψ function proposed by Huber is of the non-redescending type. This function is

$$\psi(x) = \begin{cases} x & |x| \le a \\ a\ \mathrm{sgn}(x) & |x| \ge a \end{cases} \tag{5}$$

An example of a ψ function of the redescending type is the function proposed by Hampel [6].

$$\psi(x) = \begin{cases} x & |x| \le a \\ a\ \mathrm{sgn}(x) & a \le |x| \le b \\ a\left(\dfrac{x - c\ \mathrm{sgn}(x)}{b - c}\right) & b \le |x| \le c \\ 0 & |x| \ge c \end{cases} \tag{6}$$

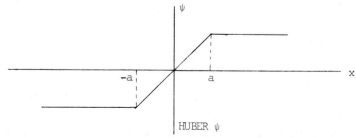

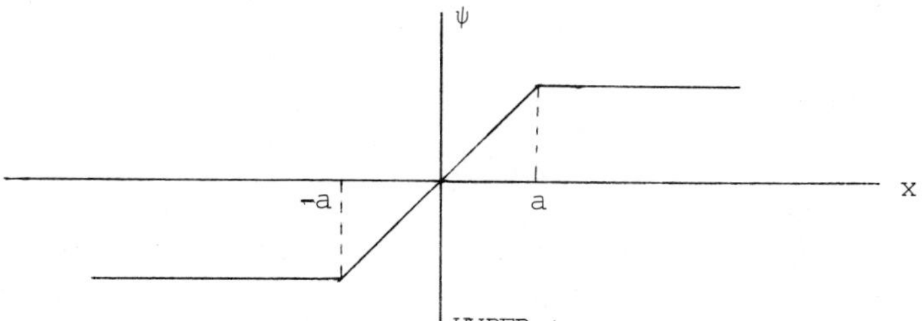

HUBER ψ

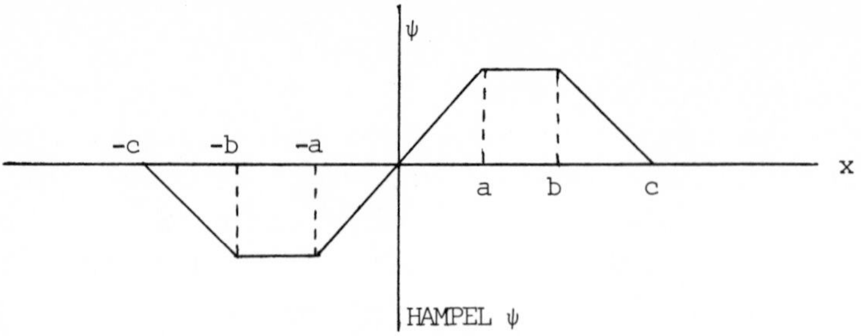

Other ψ functions have been proposed by Andrews [8], Tukey [3], and Ramsay [4]. There are also a number of other methods for robust regression, see [11].

An attractive feature of least squares regression is the ease of numerical solution. One might be inclined to think that the numerical solution for an M-estimate would in many cases be prohibitive. At worst (4) can be minimized by one of the many algorithms for minimization, e.g., the Fletcher-Powell method [10].

An iterative solution such as the Gauss-Newton method can easily be applied to minimize (4). Let $\hat{\theta}^{(k)}$ be an arbitrary point in the iteration sequence and let $r_i^{(k)}$ be the residual from this solution,

$$r_i^{(k)} = y_i - X_i\hat{\theta}^{(k)}.$$

Setting the derivative of (4) to zero and linearizing the result about $\hat{\theta}^{(k)}$ gives

$$\sum_{i=1}^{N} X_i^T \left[\psi(r_i^{(k)}/s) - \psi'(r_i^{(k)}/s)X_i(\hat{\theta}^{(k+1)} - \hat{\theta}^{(k)})/s\right] = 0. \tag{7}$$

Solving (7) for $(\hat{\theta}^{(k+1)} - \hat{\theta}^{(k)})$ yields

$$\hat{\theta}^{(k+1)} - \hat{\theta}^{(k)} = M^{-1} \sum_{i=1}^{N} \psi(r_i^{(k)}/s)X_i^T, \tag{8}$$

where

$$M = \sum_{i=1}^{N} \psi'(r_i^{(k)}/s)X_i^T X_i/s. \tag{9}$$

The iterative application of (8) and (9) results in a fairly simple method for obtaining an M-estimate. An approximate sample covariance for this estimate is given by

$$\text{cov}(\hat{\theta}) = V = \frac{1}{N-p} \sum_{i=1}^{N} \psi^2(r_i/s)M^{-1}\left(\sum_{j=1}^{N} X_j^T X_j\right)M^{-1}. \tag{10}$$

An even simpler numerical method and one which has achieved considerable popularity for obtaining M-estimates is the iterative application of weighted least squares. Setting the derivative of (4) with respect to

θ equal to zero gives

$$\sum X_i^T \psi(r_i/s) = 0, \tag{11}$$

where $r_i = y_i - X_i\hat{\theta}$. Now rewrite (11) as

$$\sum_{i=1}^{N} W_i(\hat{\theta}) X_i^T (y_i - X_i\hat{\theta}) = 0, \tag{12}$$

where

$$W_i(\hat{\theta}) = \psi(r_i/s)/(r_i/s).$$

(12) can be solved iteratively as follows. Starting at an arbitrary point $\hat{\theta}^{(k)}$ in the sequence of iterations, we replace (12) by

$$\sum_{i=1}^{N} W_i(\hat{\theta}^{(k)}) X_i^T (y_i - X_i\hat{\theta}^{(k+1)}) = 0. \tag{13}$$

Solving (13) for $\hat{\theta}^{(k+1)}$

$$\hat{\theta}^{(k+1)} = \left(\sum_{j=1}^{N} W_j(\hat{\theta}^{(k)})\, X_j^T X_j \right)^{-1} \sum_{i=1}^{N} W_i(\hat{\theta}^{(k)})\, X_i^T y_i. \tag{14}$$

Thus, we can use an ordinary weighted least squares algorithm iteratively to obtain the M-estimate.

Throughout the discussion of M-estimates we have used the dispersion measure s of the residuals $y_i - X_i\theta$ without consideration for its computation. Robust dispersion measures are often taken to be a multiple of the interquartile range or of some other range statistic of a set of residuals. A dispersion measure which seems to be most popular with those using M-estimates is the median deviation or the MAD (Median of the Absolute Deviations) estimate as it is sometimes called. The MAD estimate used above is defined by

$$s = \operatorname*{med}_i |r_i| \,/\, .6745, \tag{15}$$

where $r_i = y_i - X_i\theta$. Hampel [6] has shown that the MAD estimate is the most robust estimate of dispersion.

Both the Gauss-Newton method and the weighted least squares method for obtaining M-estimates are iterative and therefore, require a starting solution. The required closeness of the starting solution to the final solution is dependent on the application and the type of ψ function used. Quite often an ordinary unweighted least squares solution is a sufficiently good starting solution. In some cases it is necessary to use a starting solution which is more robust, see [12].

APPLICATION TO DATA PREPROCESSING

It is this application which provided our original motivation for the development and application of robust statistical methods in data reduction. There are several possible functions of data preprocessing. Ambiguities in phase measurements might be resolved by preprocessing. It might be used merely to detect outliers in the measurement data because their detection in the main processor might be considerably more difficult. Also, the main processor often requires the use of weights for each of the measurements or the main processor might require that a set of measurements be synchronized before processing. These requirements can be fulfilled by data preprocessing.

Given the time history of a particular measurement function for its entire span of observation on a trajectory, the preprocessing function divides the interval of observation into equal segments of T seconds except for a final segment either shorter or longer than T. Over each of these segments a polynomial, usually a quadratic, is fit to the measurements. Alternatively, a cubic spline might be fit to the entire span of measurement data using the end points of the T second segments as knot times. Thus, in measurement preprocessing we might model a particular measurement y over an arbitrary interval of the trajectory as

$$y_i = \theta_0 + \theta_1(t_i - \bar{t}) + \theta_2(t_i - \bar{t})^2, \ i=1,N, \tag{16}$$

where $\bar{t} = (\sum_{i=1}^{N} t_i)/N$. Using some robust M-estimate of the parameter vector $\theta = [\theta_0 \ \theta_1 \ \theta_2]$, we would minimize

$$Q = \sum_{i=1}^{N} \rho(r_i(\theta)/s), \tag{17}$$

where $r_i(\theta) = y_i - \theta_0 - \theta_1(t_i - \bar{t}) - \theta_2(t_i - \bar{t})^2$. Differentiating (17) gives the analog of the normal equations

$$\sum_{i=1}^{N} T_i^T \psi(r_i(\hat{\theta})/s) = 0, \tag{18}$$

where $T_i = [1(t_i - \bar{t})(t_i - \bar{t})^2]$. We solve (18) iteratively to obtain robust estimates $\hat{\theta}_0, \hat{\theta}_1, \hat{\theta}_2$. In the iterative solution of (18) s is taken to be the median of absolute deviations $s = \operatorname{med}_i |r_i(\hat{\theta})| \,/.6745$.

The following data set is from a real data reduction situation. The measurements are a sequence of azimuth measurements from a cinetheodolite.

OBSERVATIONS	RESIDUALS FROM ROBUST FIT	RESIDUALS FROM LEAST SQUARES FIT
-1.70987	.000012	-.157774
-1.70942	-.000004	-.000204
-1.70893	.000003	.105480
-1.70845	-.000015	.159227
-1.70793	-.000010	.161087
-1.70741	-.000021	.111021
-1.70682	.000022	.009099
-1.70626	.000019	-.144780
-1.70571	-.000010	-.350595
-1.70510	.000005	-.608277
-1.70449	.000004	-.917885
1.43777	3.141637	1.862231
1.44602	3.149243	1.456410
-1.70257	-.000007	-2.158177
1.44667	3.146558	.473139

There are three obvious outliers in the data. The residuals from an ordinary least square fit, which are given in the last column yield no information about outliers in the data. The residuals from the robust fit which were obtained using a Hampel ψ function (breakpoints 2.5, 5.0, 7.5) show exactly which observations were outliers. Outliers can be detected as those residuals r_i for which $r_i \geq ks$. The dispersion s may be saved for use in making weights for the observations in the main processing. Robust regression has been used in our trajectory measurement preprocessing program for more than a year. In this application it has been infinitely more successful than previous methods in dealing with outliers in the measurements.

Starting Solutions

Any of the numerical methods used to obtain an M-estimate requires a starting or preliminary estimate of the regression parameters θ. The starting solution is of primary importance and for some cases will determine whether or not a useable M-estimate is obtained. Robust estimation using ψ functions of the redescending type is especially sensitive to the starting solution because the solution iteration may converge to a local minimum which is relatively remote from the global minimum, if a poor starting solution is used. At best poor starting solutions require more iterations for convergence. The most obvious solution with which to start the M-estimation iteration is the unweighted least squares solution. However, since the unweighted least squares solution is highly influenced by the presence of outliers, it may not provide a suitable starting solution, $\hat{\theta}^{(0)}$. Nevertheless, least squares is often useful for starting. In some cases where the y_i are small and the components of θ are also small the starting solution $\hat{\theta}^{(0)} = 0$ may be useful. This is often the case in instrument calibration.

A good starting solution should itself be a robust estimate of the regression coefficients. Although the use of a robust starting solution may greatly increase the computing time, it will often be necessary if the two simple procedures mentioned above fail. Several robust regression methods which are suitable starting procedures for M-estimates are described in [11]. One of the simplest of these methods is an extension of the method proposed by Theil [5]. In applying this method we include a constant term θ_0 separately from the other terms in the linear model. We then apply a Gram-Schmidt orthogonalization process to the remaining independent variables. The computation of the values x'_{ij} of the orthogonal variables is given by

$$x'_{i1} = x_{i1}, \tag{19}$$

$$x'_{ij} = x_{ij} - \sum_{k=1}^{j-1} r_{jk} x'_{ik}, \tag{20}$$

$$r_{jk} = \sum_{i=1}^{N} x_{ij} x'_{ik} \bigg/ \sum_{i=1}^{N} x'^{2}_{ik}. \tag{21}$$

In terms of the orthogonal independent variables the linear model is given by

$$y_i = \theta_0 + \sum_{j=1}^{p} x'_{ij}\theta'_j + e_i, \; i=1,N. \tag{22}$$

Estimates of the regression coefficients θ'_j are obtained using our modified method of Theil by the following process.

1. $d_m(i,j) = (y_j - y_i) / (x'_{jm} - x'_{im}),\; j>i,\; i=1,\, N-1.$
2. $\delta\theta'_m = \operatorname{med}_{i,j} d_m(i,j)$
3. $\theta'_m \leftarrow \theta'_m + \delta\theta'_m$
4. $y_i \leftarrow y_i - \delta\theta'_m\, x'_{im},\; i=1,N$

(steps 1–4 for $m=1,p$)

5. Repeat steps 1 - 4 until convergence.
6. $\hat{\theta}_0 = \operatorname{med}_i y_i$

In the above $\operatorname{med}_i z_i$ means to take the median of the variables z_i over the index set i. In order to recover the original regression coefficients, it is necessary to apply the Gram-Schmidt process to the θ'_j.

$$\theta_p = \theta'_p \tag{23}$$

$$\theta_{p-i} = \theta'_{p-i} - \sum_{j=0}^{i-1} r_{p-j,p-i}\theta_{p-j}, \quad i=1,p-1. \tag{24}$$

For even moderate values of N the number of slopes $d_m(i,j)$ which must be computed is quite large. Rather than use all of these slopes we can instead work with a reduced number of slopes. One possible reduced set of slopes can be obtained letting the x'_{im} be arranged in increasing order for each m and let $N^* = [\frac{N+1}{2}]$. Thus, if N is odd $x'^*_{N^*m}$ is the median of the x'_{im}, i=1,N. We then use the slopes

$$d_m(i) = (y_{N^*+i} - y_i)/(x'^*_{N^*+i,m} - x'_{im}). \quad \begin{matrix} i=1,N^* & \text{(N even)} \\ i=1,N^*-1 & \text{(N odd)} \end{matrix}$$

These slopes are then used in step 2 of the iteration process with

$$\delta\theta_m = \operatorname*{med}_j d_m(j).$$

Another robust regression method for obtaining a starting solution for M-estimates is an application of Spearmans ρ as described in [11]. We again form a set of orthogonal independent variables x'_{im} i=1,N by applying the Gram-Schmidt process in (19) - (21). Let $R_{x_{im}}$ be the rank of x'_{im} among the x'_{jm} j=1,N and let R_{y_i} be the rank of y_i among the y_j, j=1,N. Then Spearmans ρ, a nonparametric estimate of the population correlation coefficient is defined as

$$\rho_{x_m y} = \sum_{i=1}^{N} (R_{x_{im}} - \bar{R}_{x_m})(R_{y_i} - \bar{R}_y) \Big/ \sum_{i=1}^{N} (R_{x_{im}} - \bar{R}_m)^2, \tag{25}$$

where $\bar{R}_{x_m} = \bar{R}_y = \frac{N+1}{2}$

is just the ordinary defining equation for the correlation coefficient with the variates replaced by ranks. A more useful definition of $\rho_{x_m y}$ for computing is

$$\rho_{x_m y} = 1 - \frac{6}{N(N^2-1)} \sum_{i=1}^{N} d_i^{\,2}, \tag{26}$$

where d_i is the rank difference

$$d_i = R_{y_i} - R_{x_{im}}.$$

In an orthogonal regression model the estimates of the regression coefficients may be written as

$$\hat{\theta}'_m = \hat{\rho}_{x_m y}\, \hat{\sigma}_y / \hat{\sigma}_{x_m}, \tag{27}$$

where $\hat{\rho}_{x_m y}$, $\hat{\sigma}_y$, $\hat{\sigma}_{x_m}$ are the usual sample correlation coefficient and standard deviations. An obvious method of obtaining a robust estimate of θ'_m is to replace $\hat{\rho}_{x_m y}$, $\hat{\sigma}_y$, $\hat{\sigma}_{x_m}$ in (27) by nonparametric estimates of these quantities. Thus, we replace $\hat{\rho}_{x_m y}$ by Spearmans ρ and replace $\hat{\sigma}_y$ by

$$\hat{\sigma}_y = \operatorname*{med}_i |y_i - y^*| \Big/ .6745, \tag{28}$$

where $y^* = \operatorname*{med}_i y_i$. We could also replace $\hat{\sigma}_{x_m}$ by an estimate similar to (28) but in most cases $\hat{\sigma}^2_{x_m} = \frac{1}{N-1} \sum_{i=1}^{N} (x'_{im} - \bar{x}'_m)^2$ is sufficient. The process is used iteratively to improve the estimate of θ'_m. The procedure is implemented by the following steps.

1. $R_{x_{im}} = \text{rank } x'_{im}$

   $\hat{\sigma}^2_{x_m} = \frac{1}{N-1} \sum_{i=1}^{N} (x'_{im} - \bar{x}'_m)^2$ $\Big\}$ m=1,p

2. $R_{y_i} = \text{rank } y_i$

   $y^* = \operatorname*{med}_i y_i$

   $\hat{\sigma}_y = \operatorname*{med}_i |y_i - y^*| \Big/ .6745$

3. $d_i = R_{y_i} - R_{x_{im}}$

   $\delta\rho_m = 1 - \frac{6}{N(N^2-1)} \sum_{i=1}^{N} d_i^2$

   $\delta\theta'_m = \delta\rho_m \hat{\sigma}_y / \hat{\sigma}_{x_m}$

   $\theta'_m \leftarrow \theta'_m + \delta\theta'_m$

   $y_i \leftarrow y_i - \delta\theta'_m x'_{im}$ i=1,N

   (steps 2–3: $\Big\}$ m=1,p)

4. Repeat steps 2 - 3 until convergence.
5. $\hat{\theta}_0 = \operatorname*{med}_i y_i$

As before we must apply the Gram-Schmidt process to the θ'_m in order to recover the original regression coefficients. The above process can also be implemented using Kendalls τ rather than Spearmans ρ as described in [11].

A third method for obtaining a robust starting solution is the orthogonal Brown-Mood method. This is a variation of the Brown-Mood method [9] which uses orthogonal independent variables. Let

x'_{im}, m=1,p, i=1,N be a set of orthogonal independent variables obtained by applying the Gram-Schmidt process. Let x^*_m be the median of the x'_{im} i=1,N. The Brown-Mood method is iterative so let $\hat{\theta}^{(k)}$ be some estimate in the iteration sequence and let $r_i^{(k)}$ be the residuals

$$r_i^{(k)} = y_i - \sum_{m=1}^{p} x'_{im}\hat{\theta}_m^{(k)}. \tag{29}$$

The Brown-Mood method computes corrections $\delta\theta'_m$ to $\hat{\theta}^{(k)}$ by

$$\delta\theta'_m = (r_i^{(k)^+} - r_i^{(k)^-}) \; / \; (x_m^+ - x_m^-), \tag{30}$$

where

$$x_m^+ = \operatorname*{med}_{i\epsilon I_U} x'_{im}; \; I_U = \{i|x'_{im} > x^*_m\}, \tag{31}$$

$$x_m^- = \operatorname*{med}_{i\epsilon I_L} x'_{im}; \; I_L = \{i|x'_{im} \leq x^*_m\}, \tag{32}$$

$$r_i^{(k)^+} = \operatorname*{med}_{i\epsilon I_U} r_i^{(k)}; \; r_i^{(k)^-} = \operatorname*{med}_{i\epsilon I_L} r_i^{(k)}. \tag{33,34}$$

The estimates are updated by $\hat{\theta}'^{(k+1)}_m \leftarrow \hat{\theta}'^{(k)} + \delta\theta'_m$ and the above procedure is iterated to convergence. Finally, the estimate of θ_0 is obtained from

$$\hat{\theta}_0 = \operatorname*{med}_i r_i^{(k)}$$

The orthogonal Brown-Mood method is implemented by the following steps (starting with $\hat{\theta}^{(0)} = 0$).

$$\left.\begin{array}{ll}
1. & x^*_m = \operatorname*{med}_i x'_{im} \\[1ex]
2. & x_m^+ = \operatorname*{med}_{i\epsilon I_U} x'_{im}; \quad x_m^- = \operatorname*{med}_{i\epsilon I_L} x'_{im} \\[1ex]
3. & y_i^+ = \operatorname*{med}_{i\epsilon I_U} y_i \; ; \quad y_i^- = \operatorname*{med}_{i\epsilon I_L} y_i \\[1ex]
4. & \delta\theta'_m = (y_i^+ - y_i^-) \; / \; (x_m^+ - x_m^-) \\[1ex]
 & \theta'_m \leftarrow \theta'_m + \delta\theta'_m \\[1ex]
5. & y_i \leftarrow y_i - \delta\theta'_m \, x'_{im} \quad i=1,N \\[1ex]
6. & \text{Repeat steps 3 - 5 until convergence.}
\end{array}\right\} \; m=1,p$$

$$7. \quad \theta_0 = \operatorname*{med}_i y_i .$$

INSTRUMENT CALIBRATION

Surveyed targets are used for calibrating, i.e., estimating the coefficients in an error model, for radars, cinetheodolites or laser trackers. Suppose for example we have M surveyed targets for a laser tracker. Let R_{sj}, A_{sj}, E_{sj} be the surveyed range, azimuth, and elevation for the jth target. Suppose that multiple observations of the targets are available so that we have N_j observations for the jth target. Denote these range, azimuth and elevation observations by R_{ij}, A_{ij}, and E_{ij}, $i=1,N_j$, $j=1,M$. Let

$$\Delta R_{ij} = R_{ij} - R_{sj} = r_j^T\theta + \text{(random error)}$$
$$\Delta A_{ij} = A_{ij} - A_{sj} = a_j^T\theta + \text{(random error)}$$
$$\Delta E_{ij} = E_{ij} - E_{sj} = e_j^T\theta + \text{(random error)}$$

where θ is an unknown parameter vector and r_j, a_j, and e_j are known vectors. A common model for r_j, a_j, and e_j is given by

$$r_j^T\theta = \theta_1 + \theta_2 R_{sj}, \tag{35}$$
$$a_j^T\theta = \theta_3 - \theta_4 \tan E_{sj}\cos A_{sj} - \theta_5 \tan E_{sj}\sin A_{sj} - \theta_6/\cos E_{sj}, \tag{36}$$
$$e_j^T\theta = \theta_7 + \theta_4 \sin A_{sj} - \theta_5 \cos A_{sj}. \tag{37}$$

The usual least squares estimate of the parameter vector θ would minimize

$$\sum_{j=1}^{M}\sum_{i=1}^{N_j}\left[(\Delta R_{ij} - r_j^T\theta)^2 + (\Delta A_{ij} - a_j^T\theta)^2 + (\Delta E_{ij} - e_j^T\theta)^2\right] \tag{38}$$

An M-estimate alternative to least squares would minimize

$$\sum_{j=1}^{M}\sum_{i=1}^{N_j}\left[\rho\left((\Delta R_{ij}-r_j^T\theta)/s_r\right) + \rho\left((\Delta A_{ij}-a_j^T\theta)/s_a\right) + \rho\left((\Delta E_{ij}-e_j^T\theta)/s_e\right)\right] \tag{39}$$

Differentiating (39) gives the analog to the normal equations

$$\sum_{j=1}^{M}\sum_{i=1}^{N_j}\left[(r_j/s_r)\,\rho\left((\Delta R_{ij}-r_j^T\theta)/s_r\right) + (a_j/s_a)\,\rho\left((\Delta A_{ij}-a_j^T\theta)/s_a\right) + (e_j/s_e)\,\rho\left((\Delta E_{ij}-e_j^T\theta)/s_e\right)\right] = 0. \tag{40}$$

An iterative solution of (40) with

$$s_r = \operatorname*{med}_{i,j}\left|\Delta R_{ij}-r_j^T\hat\theta^{(k)}\right| \Big/ .6745; \qquad s_a = \operatorname*{med}_{i,j}\left|\Delta A_{ij}-a_j^T\hat\theta^{(k)}\right| \Big/ .6745;$$
$$s_e = \operatorname*{med}_{i,j}\left|\Delta E_{ij}-e_j^T\hat\theta^{(k)}\right| \Big/ .6745;$$

gives a robust estimate $\hat\theta$. Since the elements of the parameter vector are usually small, the elements of the starting solution $\theta^{\{0\}}$ may be set to zero except for $\theta_1^{\{0\}}$, $\theta_3^{\{0\}}$, and $\theta_7^{\{0\}}$ which can be set to the medians of R_{ij}, A_{ij}, and E_{ij}, respectively.

The following example illustrates the application of M-estimates to the calibration of a laser tracker using real field data. The laser tracker is calibrated by using range, azimuth and elevation observations from eight reflective targets arranged in a circular pattern around the tracker at a range of about 2500 feet. We use the error model given in (35) - (37). Since the elevations of the eight calibration targets are approximately equal, it is obviously impossible to estimate θ_6 in (36) without additional observations. In order to provide these additional observations we observe the same calibration targets but with the tracker "dumped", i.e., with an azimuth of approximately $A_{si} + 180°$ and an elevation of approximately $E_{si} - 180°$. These additional observations are called dumped readings and are treated as additional calibration targets. Also, we can see from (35) that we will be unable to estimate θ_2 using the eight calibration targets since the ranges to all targets are approximately equal. In order to estimate θ_2 we observe four additional calibration targets with ranges varying from 20000 feet to 60000 feet. Approximately 250 observations are available for each of the remaining target boards.

A Hampel ψ function which was defined in (6) was used for this example. The parameters or break points of the Hampel ψ in this example are $a = 2.5$, $b = 5.0$, $c = 7.5$. The results of this robust calibration are summarized in the table on the following page by tabulating the number of residuals for each target lying in each region of the Hampel ψ. We show only the positive side of the ψ function with the number of residuals in each region being the sum of the numbers of residuals in the positive and corresponding negative side of the ψ function. The first eight target boards are at 2500 ft. circularly about the tracker. Targets 9-12 are the long range calibration targets. Targets 13-20 are the dumped readings of the first eight targets. From the table it can be seen that most of the observations from several target boards are outliers, particularly for the dumped readings. A least squares estimation of the calibration parameters failed miserably on this example. Although this example is extreme for this application, having about 22% contamination by outliers, it well illustrates the power of the M-estimate method in dealing with outliers.

N-STATION CINETHEODOLITE SOLUTION

The N-station Cine solution is a standard problem in data reduction. In this situation we are given azimuth observations $a_\alpha(t_i)$ and elevation observations $e_\alpha(t_i)$, $\alpha = 1, N_i$, at each time point t_i along a trajectory. From these N_i cines we must estimate the cartesian positions $x(t_i)$, $y(t_i)$,

NUMBER OF RESIDUALS

Target Pole #	$<2.5s_r$	$(2.5s_r, 5s_r)$	$(5s_r, 7.5s_r)$	$>7.5s_r$	$<2.5s_a$	$(2.5s_a, 5s_a)$	$(5s_a, 7.5s_a)$	$>7.5s_a$	$<2.5s_e$	$(2.5s_e, 5s_e)$	$(5s_e, 7.5s_e)$	$>7.5s_e$
1	230	0	0	3	230	0	0	3	230	0	0	3
2	252	0	0	0	251	0	0	1	252	0	0	0
3	237	0	0	0	237	0	0	0	237	0	0	0
4	270	0	0	0	270	0	0	0	270	0	0	0
5	241	1	0	0	242	0	0	0	242	0	0	0
6	242	0	0	0	242	0	0	0	242	0	0	0
7	237	0	0	0	237	0	0	0	237	0	0	0
8	215	0	0	9	215	0	0	9	222	2	0	0
9	9	0	0	241	7	2	0	241	193	57	0	0
10	269	15	0	0	243	40	1	0	284	0	0	0
11	250	1	0	0	245	6	0	0	251	0	0	0
12	161	2	0	118	127	35	55	64	217	0	0	64
13	118	103	0	5	224	0	0	2	38	186	0	2
14	135	86	25	4	248	0	0	2	234	13	0	3
15	2	0	0	0	2	0	0	0	2	0	0	0
16	126	59	9	33	221	0	0	6	21	200	0	6
17	138	96	20	39	248	0	0	45	248	44	1	0
18	2	0	0	0	0	0	0	2	0	2	0	0
19	137	71	9	17	0	0	0	234	8	226	0	0
20	81	86	18	111	0	0	0	296	296	0	0	0
	RANGE				AZIMUTH				ELEVATION			

DISTRIBUTION OF CALIBRATION RESIDUALS

$z(t_i)$, at each time point. The observations are $a_\alpha(t_i) = A_\alpha(\bar{x}_i)$ + error and $e_\alpha(t_i) = E_\alpha(\bar{x}_i)$ + error. The measurement functions $A_\alpha(\bar{x}_i)$ and $E_\alpha(\bar{x}_i)$ are functions of the position vector $\bar{x}_i = [x(t_i)y(t_i)z(t_i)]$. These measurement functions are given by

$$A_\alpha(\bar{x}_i) = \tan^{-1}\{(x(t_i) - x_\alpha)/(y(t_i) - y_\alpha)\}, \tag{41}$$

$$E_\alpha(\bar{x}_i) = \tan^{-1}\{(z(t_i) - z_\alpha)/[(x(t_i)-x_\alpha)^2+(y(t_i)-y_\alpha)^2]^{1/2}\}. \tag{42}$$

where $(x_\alpha, y_\alpha, z_\alpha)$ is the cartesian position of the αth cine. The usual least square problem to estimate the position $x(t_i)$, $y(t_i)$, $z(t_i)$ is non-linear. Thus, the robust estimation of these quantities will be nonlinear both because the objective function for the robust estimation problem is non-quadratic and because the measurement model is a nonlinear function of the parameters to be estimated. The usual least squares solution would minimize

$$\sum_{\alpha=1}^{N_i}\left[\left(a_\alpha(t_i) - A_\alpha(\bar{x}_i)\right)^2\cos^2 e_\alpha(t_i) + \left(e_\alpha(t_i) - E_\alpha(\bar{x}_i)\right)^2\right]. \tag{43}$$

An M-estimate of the position vector $\bar{x}_i$ would minimize

$$\sum_{\alpha=1}^{N_i}\left[\rho\left((a_\alpha(t_i)-A_\alpha(t_i))/s_a\right)\cos^2 e_\alpha(t_i)+\rho\left((e_\alpha(t_i)-E_\alpha(\bar{x}_i))/s_e\right)\right]. \tag{44}$$

Differentiating (44) gives

$$\sum_{\alpha=1}^{N_i}[(\partial A_\alpha(\bar{x}_i)/\partial\bar{x}_i)\psi(r_a(\alpha)/s_a)\cos^2 e_\alpha(t_i)/s_a + (\partial E_\alpha(\bar{x}_i)/\partial\bar{x}_i)\psi(r_e(\alpha)/s_e)/s_e]=0, \tag{45}$$

where $r_a(\alpha) = a_\alpha(t_i) - A_\alpha(\bar{x}_i)$; $r_e(\alpha) = e_\alpha(t_i) - E_\alpha(\bar{x}_i)$.

(45) can be rewritten as

$$\sum_{\alpha=1}^{N_i} C_\alpha^T(\bar{x}_i)\psi(r_a(\alpha)/s_a, r_e(\alpha)/s_e)=0, \tag{46}$$

where

$$C_\alpha^T(\bar{x}_i) = [(\partial A_\alpha(\bar{x}_i)/\partial\bar{x}_i)/s_a(\partial E_\alpha(\bar{x}_i)/\partial\bar{x}_i)/s_e] \tag{47}$$

is a 3 x 2 matrix and

$$\psi^T(r_a(\alpha)/s_a, r_e(\alpha)/s_e) = [\psi(r_a(\alpha)/s_a)\cos^2 e_\alpha(t_i)\psi(r_e(\alpha)/s_e)]. \tag{48}$$

An iterative solution of (46) with $s_a = \operatorname{med}_\alpha |r_a(\alpha)| / .6745$, $s_e = \operatorname{med}_\alpha |r_e(\alpha)| / .6745$ gives a robust estimate of the parameter vector $\bar{x}_i$.

As an example of robust estimation applied to a cinetheodolite solution consider the following situation which is rather extreme but sometimes occurs. A missile is fired at a drone and cinetheodolites are observing both the missile and drone. It is required to provide a cine derived trajectory on both the missile and the drone. Due to an inadvertent clerical error, one of the cines which was actually observing the missile was erroneously listed as observing the drone. Obviously, when doing a least squares solution to obtain the drone trajectory, the azimuth and elevations from one cine will be gross outliers and may destroy the least squares solution for the drone position coordinates. A single point example of this situation is furnished by the actual cine data given below

Cine	Obs. Azimuth	Obs. Elevation
1	.568106	.338886
2	-.626010	.122620
3	-2.665036	.359168
4	1.926249	.327177

Cine 2 is the one which is actually tracking the missile rather than the drone. Obviously, as in most situations which are nonlinear, there is no way of distinguishing the outliers by inspecting the observations. As always in robust estimation a preliminary solution is required to start the iteration. Let $(x_\alpha, y_\alpha, z_\alpha)$ be a position solution obtained from the αth pair of cines. In this example we have six possible pairs of cines so that $\alpha = 1, 6$. We then start the iteration with (x^o, y^o, z^o) where $x^o = \underset{\alpha=1,6}{\text{med}}\ x_\alpha$, $y^o = \underset{\alpha=1,6}{\text{med}}\ y_\alpha$, $z^o = \underset{\alpha=1,6}{\text{med}}\ z_\alpha$. For the example, the median guess solution is $x^o = -45147.9$ ft., $y^o = 87423.8$ ft., $z^o = 11117.3$ ft. After five iterations the sequence has converged to the solution $x = 32964.8$ ft., $y = 87425.2$ ft., $z = 11114.9$ ft. The residuals from the final solution are

RESIDUALS

Cine	Azimuth	Elevation
1	.000008	-.000064
2	-.242553	.011513
3	.000022	.000081
4	.000057	-.000019

Thus, the robust solution using the Hampel ψ with breakpoints of 3, 6, 9, correctly identified the outliers. Let us carry this example farther. Suppose we have no observations from Cine 1, i.e., we have data from only three cines, one of which is bad. In this case our starting solution turns out to be x^o = 45147.9, y^o = 87424.1, z^o = 11120.2. After four iterations the solution has converged to x = -32966, y = 87424.6, z = 11115.3. Thus, we are again able to correctly identify the bad cine. Now suppose we have data from cines 1, 2, 3. In this case the initial guess solution is x^o = 45147.9, y^o = 67033.9, z^o = 11118.9. After ten iterations the solution is x = -35023.9, y = 84462.1, z = 11004.1. The solution eventually converges to the correct value, but slowly. A third possibility to have data from only three cines is observations from cines 1, 2, 4. In this case the guess solution is x^o = -46454.3, y^o = 87548.3, z^o = 7262.7. After three iterations the solution has converged to x = -35392.6, y = 86464.3, z = 1044.8. Thus, in this case the iteration has converged to the wrong solution. In the last two cases where the solution converged very slowly and converged to the wrong solution, the starting solution was too far from the correct solution. If a sufficiently good start had been provided the solution would have converged correctly in a few iterations. If the number of cines were great enough in comparison to the number of bad cines, using the median of the solutions obtained from the cine pairs provides an acceptable starting solution. Unfortunately, the number of cines is often no more than three or four. In the case of three cines, the use of a starting solution predicted from preceding points might be a desirable procedure. If preprocessing had been used on the cine data most, if not all, of the outliers of the spike variety in the cine data would have been detected before attempting a solution. Thus, robust estimation in the solution has only to contend with detecting badly biased cines. In any situation with three or more cines with one bad cine, the robust solution will usually provide a better solution than the usual least square procedure. A strategy for choosing a good starting solution needs to be developed. A robust N-station radar solution is developed along the same lines as a robust cine solution. In the radar case a starting solution for the iteration is somewhat easier to obtain.

APPLICATION TO RECURSIVE FILTERING

Very little development has been done on the application of robust statistical techniques to filtering. The most significant effort known to the author is given in the paper of Masreliez and Martin [7]. Their development of robustifying the Kalman filter is quite complex and will not be considered here. It is a simple matter to specify a form for an approximate M-filter and its covariance.

Suppose we wish to estimate the state x(n) of the linear dynamic model described by the state equation.

$$x(n + 1) = \Phi(n + 1, n)\ x(n) + u(n), \tag{49}$$

where $\Phi(n + 1, n)$ is an mxm state transition matrix and u(n) is an m-vector of state noise with covariance Q_n. Suppose we are also given scalar observations Z(n) of the state specified by

$$Z(n) = Hx(n) + v(n), \tag{50}$$

where H is a 1xm matrix of constants and v(n) is observation error. By analogy with the least squares filter derivation we minimize

$$\sum_{i=1}^{N} \rho\left((Z(i) - Hx(i))/s_i\right) + u^T(i)\ Q_i^{-1}\ u(i)/2, \tag{51}$$

subject to the constraints

$$x(i + 1) - \Phi x(i) - u(i) = 0,\ i=1,n-1. \tag{52}$$

Minimizing (51) leads to the approximate filter equations

$$\hat{x}(n+1/n+1)=\hat{x}(n+1/n)+P_{n+1}H^T\psi\left((Z(n+1)-H\hat{x}(n+1/n))/s_{n+1}\right)\Big/s_{n+1}, \tag{53}$$

$$\hat{x}(n+1/n) = \Phi\hat{x}(n). \tag{54}$$

The approximate covariances of $\hat{x}(n+1/n+1)$ is given by

$$P_{n+1}^{-1} = P_{n+1/n}^{-1}+H^T H\psi'\left((Z(n+1)-H\hat{x}(n+1/n))/s_{n+1}\right)\Big/s_{n+1}^2, \tag{55}$$

where

$$P_{n+1/n} = \Phi P_n \Phi^T + Q_n. \tag{56}$$

This robust filter is certainly easy to implement and anyone who has done much recursive filtering of data on a computer has probably implemented such a filter with the following ψ function,

$$\psi(x) = \begin{cases} x & |x| \le k \\ 0 & |x| \ge k \end{cases}$$

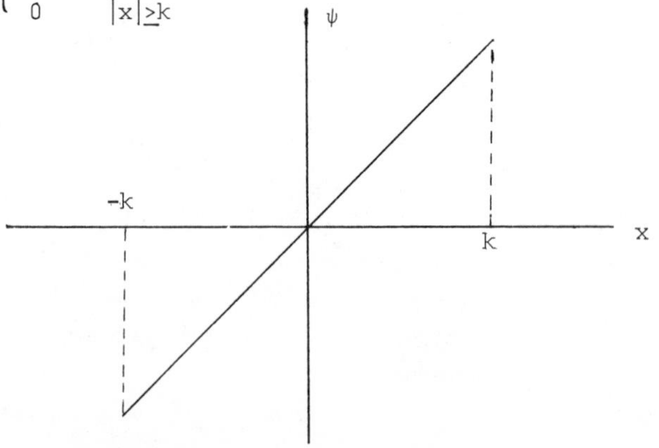

i.e., we process observations only if the predicted residuals are within $\pm k\hat{\sigma}$ where $\hat{\sigma}$ is an estimate of the measurement noise standard deviation. Thus, robust filtering presents nothing new as far as filter implementation is concerned, but we are now in a position to possibly improve our robust filtering by borrowing some ψ functions and other concepts which have proved very useful in robust regression.

FUTURE EFFORT

Efforts to apply robust methods to various trajectory estimation problems at WSMR are continuing. Of immediate interest is to determine the advantages to be derived from applying robust filtering and smoothing methods to aircraft and missile tracking data. This work will be reported on in the near future. Also, we are currently working on the application of another robust regression method which on the basis of a few tests seems to offer considerable promise. This method, which we shall call the SAD (Sum of Absolute Deviations) method, is equivalent to the method reported on by Hettmansperger and McKean [13] in which they used the median scores to derive a nonparametric regression scheme. The SAD estimate of the regression coefficients θ_j in the linear model (1) minimizes

$$\sum_{i=1}^{N} |r_i - r^*| \tag{57}$$

where the r_i are the residuals

$$r_i = y_i - \sum_{j=1}^{P} \theta_j x_{ij} \tag{58}$$

and $r^* = \underset{i}{\text{med}}\ r_i$. Besides offering a robust estimate of the regression coefficients, the SAD estimate has the advantage that the function being minimized is convex as shown by Jaekel [14]. Another attractive feature of this method is that, like the M-estimates, weighted least squares can be applied iteratively to obtain the SAD estimate.

REFERENCES

1. Huber, Peter J., (1973). Robust Regression: Asymptotics, Conjectures, and Monte Carlo, Ann. Statist., 1, 799-821.
2. Huber, Peter J., (1972). Robust Statistics: A Review, Ann. Math. Statist., 43, 1041-1067.
3. Tukey, John W., and Beaton, Albert E., (May 1974). The Fitting of Power Series, Meaning Polynomials, Illustrated on Band-Spectroscopic Data, Technometrics, 16, 147-192.

4. Ramsay, J. O., (Sept. 1977). A Comparative Study of Several Robust Estimates of Slope, Intercept, and Scale in Linear Regression, J. Amer. Statist. Assoc., 72, 608-615.

5. Theil, H., (1950). A Rank-Invariant Method of Linear and Polynomial Regression Analysis, Indag. Math., 12, 85-91, 173-177, 467-482.

6. Hampel, Frank R., (June 1974). The Influence Curve and its Role in Robust Estimation, J. Amer. Statist. Assoc., 69, 383-393.

7. Masreliez, C. Johan and Martin, R. Douglas, (June 1977). Robust Bayesian Estimation for the Linear Model and Robustifying the Kalman Filter, IEEE Trans. Automatic Control, AC-22, 361-371.

8. Andrews, D. F., (Nov. 1974). A Robust Method for Multiple Linear Regression, Technometrics, 16, 523-531.

9. Mood, A. M., (1950). Introduction to the Theory of Statistics, McGraw-Hill, New York.

10. Fletcher, K., and Powell, M. J. D., (1963). A Rapidly Convergent Descent Method for Minimization, Computer J., 6, 163-168.

11. Agee, W. S. and Turner, R. H., (1978). Robust Regression: Some New Methods and Improvements of Old Methods, Technical Report, White Sands Missile Range.

12. Agee, W. S. and Turner, R. H., (1978). Robust Regression: Computational Methods for M-estimates, Technical Report, White Sands Missile Range.

13. Hettmansperger, Thomas P. and McKean, Joseph W., (1977). A Robust Alternative Based on Ranks to Least Squares in Analyzing Linear Models, Technometrics, 19, 275-284.

14. Jaekel, L. A., (1974). Estimating Regression Coefficients by Minimizing the Dispersion of the Residuals, Ann. Math. Statist., 43, 1449-58.

Mathematical Services Branch
Analysis and Computation Division
National Range Operations Directorate
US Army White Sands Missile Range
White Sands Missile Range, NM 88002

Tests for Censoring of Extreme Values (Especially) When Population Distributions Are Incompletely Defined

N. L. Johnson

1. INTRODUCTION

This paper contains a review of a series of studies by the author ([1]-[13]) spread over some 15 years, though with by far the greatest concentration over the last five years or so ([6]-[13]). Although an account of current lines of work is included, emphasis is on completed research.

The common feature of this work is concern with the problem of deciding whether a given set of data represents a complete random sample, or only the remainder of a sample after some form of censoring (removal of specified order statistics) has been applied. Within this general framework, many variations are possible, some of which will be discussed. We will, however, always observe the overriding constraints that sampling is assumed to be random from effectively infinite populations, and that the measurement(s) can be regarded as observed value(s) of continuous random variables.

It is clearly impossible to construct useful procedures for deciding whether *any* set of observed values represents a complete random sample (the alternatives being that it has been censored in some way) without specifying something more about the population distribution than simply requiring it to be continuous. It is always possible to construct popu-

† I would like to express my gratitude, for assistance in computing, to Dr. Helen T. Bhattacharyya (Table 4), Ms. Anne Colosi (Table 6) and to Dr. Kerry L. Lee (Tables 2 and 3).

ISBN 0-12-438150-2

lation distributions which would indicate acceptance of completeness for a given set of values, and other distributions which would lead to rejection of the same hypothesis.

If the population distribution is known *exactly*, it is possible to construct test procedures which are optimal in certain senses. This was done in [2]-[5]; the results are described briefly in Section 2. Subsequent work was aimed at:

(i) assessing possible effects of inaccuracy in the population distribution assumed

(ii) investigating whether useful procedures could be constructed on the basis of partial knowledge of population distributions.

Topic (i) is discussed, again rather briefly, in Section 3; topic (ii) is discussed, at somewhat greater length, in Section 4. The "partial knowledge" is first supposed to be in the form of values from a *complete* random sample from the population; this requirement is then relaxed by supposing only that some (possibly a specified minimum number) out of a number of sets of values, correspond to complete random samples, it not being known *which* sets are uncensored.

In an early paper [2], general types of censoring were considered. In order to avoid too lengthy discussions, later papers discussed only censoring of extreme values from a complete random sample. This limitation will also be observed here. We will distinguish

(a) censoring from above (exclusion of s_r greatest values), or censoring from below (exclusion of s_0 least values)

(b) symmetrical censoring ($s_0 = s_r = s$ - equal from above and below)

(c) general censoring of extreme values (s_0, s_r unknown).

Sections 2-4 are concerned only with univariate problems. In Section 5 there is some discussion of multivariate problems. Of course the concept of "censoring of extreme values" needs to be modified when dealing with multivariate data, though one can have such censoring in respect to one of the variables. The variety of possible problems is naturally much richer than for univariate data.

2. SOME OPTIMAL TEST PROCEDURES

We suppose that we have r observed values, giving order statistics $X_1 \le X_2 \le \ldots \le X_r$. It is assumed that these are part (or all) of a random sample from an infinite continuous population, in which the probability density function (PDF) of the measured variate is f(x), and the cumulative density function (CDF) is

$$F(x) = \Pr[X \le x] = \int_{-\infty}^{x} f(t)dt \ . \tag{1}$$

For convenience we will use the notation

$$Y_i = F(X_i) \quad (i=1,\ldots,r). \tag{2}$$

As explained in Section 1, we restrict ourselves to the possibilities that there was an original complete random sample of size n, from which the s_0 least and s_r greatest values have been removed. Clearly

$$n = r + s_0 + s_r \tag{3}$$

and, if there has been no censoring, $s_0 = s_r = 0$.

Censoring from above corresponds to $s_0 = 0$, $s_r > 0$

Censoring from below corresponds to $s_0 > 0$, $s_r = 0$

Symmetrical censoring corresponds to $s_0 = s_r > 0$.

We use H_{s_0,s_r} to denote the hypothesis that s_0, s_r have specified values: $H_{0,0}$ is the hypothesis of no censoring.

The likelihood of the observed values $(X_1 \le X_2 \le \ldots \le X_r)$ is

$$L(H_{s_0,s_r}) = \frac{(r+s_0+s_r)!}{s_0!s_r!}\{F(X_1)\}^{s_0}\{1-F(X_r)\}^{s_r}\prod_{i=1}^{r} f(X_i). \tag{4}$$

Supposing, for the moment, that we wish to discriminate between the two hypotheses

$$H' \equiv H_{s_0',s_r'} \quad \text{and} \quad H \equiv H_{s_0,s_r},$$

we would use the likelihood ratio

$$\frac{L(H)}{L(H')} = \frac{L(H_{s_0,s_r})}{L(H_{s_0',s_r'})} = C\{F(X_1)\}^{s_0-s_0'}\{1-F(X_r)\}^{s_r-s_r'}$$

$$= CY_1^{s_0-s_0'}(1-Y_r)^{s_r-s_r'} \tag{5}$$

where the constant C depends on r, s_0', s_0, s_r' and s_r but not on the Y's. A procedure of type

$$\text{"Accept } H'(H) \text{ if } Y_1^{s_0-s_0'}(1-Y_r)^{s_r-s_r'} > (<) K \tag{6}$$

is optimal in the sense that if the probability of incorrect decision when one of the two hypotheses (H',H) is valid, the probability of incorrect decision, when the other is valid, is minimized. The value of K can be chosen arbitrarily, to control one of these two probabilities of incorrect decision. The *form* of the inequality (6) depends only on the ratio $(s_r-s_r')/(s_0-s_0')$, since (6) can be written

$$Y_1(1-Y_r)^{(s_r-s'_r)/(s_0-s'_0)} > (<)\ K_1 \ . \tag{7}$$

(If $s_0 = s'_0$, the inequality is just $1 - Y_r > (<)\ K_1$.) The value K_1, of course, depends on the value specified for the probability of incorrect decision and on r, s'_0, and s'_r (or r, s_0 and s_r) as well as on $(s_r-s'_r)/(s_0-s'_0)$.

If $s'_0 = s'_r = 0$, so that $H' \equiv H_{0,0}$ (no censoring) then the test becomes

$$\text{Accept H (i.e. reject } H_{0,0}) \text{ if } \ Y_1(1-Y_r)^{s_r/s_0} > K_1 . \tag{8}$$

If the value of K_1 is chosen to make the probability of rejection of $H_{0,0}$, when it is valid (significance level) equal to a specified value α, then K_1 depends only on α, r and s_r/s_0 (since it is known that $s'_0 = s'_r = 0$).

The *same* optimal test is obtained for all H (alternative hypotheses H_{s_0,s_r}) for which s_r/s_0 has the same value, θ, say. This test is therefore "uniformly most powerful" with respect to this set of alternative hypotheses. In particular there are uniformly most powerful tests of $H_{0,0}$ with respect to

(i) censoring from below ($s_r/s_0 = 0$): critical (rejection) region
$$Y_1 > K_1 \tag{9.1}$$
(ii) censoring from above ($s_r/s_0 = \infty$): cricital (rejection) region
$$1-Y_r > K_1 \tag{9.2}$$
(iii) symmetrical censoring ($s_r/s_0 = 1$): critical (rejection) region
$$Y_1(1-Y_r) > K_2 \ . \tag{10}$$

(The value of K_1 in (9.1) and (9.2) is simply $1 - \alpha^{1/r}$.)

If the possible values of s_0, s_r are to be completely general, we have to use some other approach. In [4], S.N. Roy's union-intersection principle approach was used, combined with a number of approximations, to yield the critical region

$$Y_1 + (1-Y_r) > K_3 \ . \tag{11}$$

Power properties of these three kinds of test ((9), (10), (11)) were discussed in [5]. The results were generally as might be expected. The "general purpose" test (11), is not so good as the optimal test ((9), (10)) when the latter is used in an appropriate situation, but it is better than any of the optimal tests used in some other situations. Table 1 gives a few numerical values. Among other interesting results in [5], we note

(a) if r tends to infinity, s_0 and s_r remaining constant, the power of each of the tests (9) - (11) increases, but tends to a limit less than 1 (shown in the $r = \infty$ rows in Table 1.

(b) If r is kept fixed and s_0, s_r tend to infinity with $s_r/s_0 \to \theta$, the power of tests (9.1) and (9.2) tend to 0 or 1 depending on the values of θ and α (the significance level). For (9.1) the power tends to 0 (1) if $\alpha < (>) \{\theta(1+\theta)^{-1}\}^r$; for (9.2) the situation is reversed.

Test Criterion	r	s_0+s_r =	2			6			10	
		(s_0,s_r)=(0,2)	(1,1)	(2,0)	(0,6)	(3,3)	(6,0)	(0,10)	(5,5)	(10,0)
(9.1) Y_1	4	.011	.086	.294	.001	.131	.780	<.0005	.157	.955
	∞	.050	.200	.424	.050	.648	.967	.050	.996	.999
(9.2) $1-Y_r$	4	.294	.086	.011	.780	.131	.001	.955	.157	<.0005
	∞	.424	.200	.050	.967	.648	.050	.999	.996	.050
(10) $Y_1(1-Y_r)$	4	.124	.206	.124	.120	.594	.120	.080	.841	.080
	∞	.239	.339	.239	.533	.925	.533	.644	.998	.644
(11) $Y_1+(1-Y_r)$	4		.167			.470			.716	
	∞		.303			.892			.996	

TABLE 1. Comparison of Powers. (α = 0.05)

If we have available two, or more sets of observed values, and know (somehow) that either all, or none of them have been censored (the censoring system being the same, in each case) the tests (9) - (11) can be extended in a straightforward manner [6]. There are quite remarkable increases in power, even when only two or three sets are available. However, it must be admitted that the assumption that, if there is censoring, the censoring system is the same for each set is the same, is rather stringent. On the other hand, the theory can be adapted (at the expense of some formal elaboration) to allow for varying systems of censoring having certain elements in common (e.g. with s_r/s_0 between certain fixed values).

3. ROBUSTNESS OF TEST PROCEDURES

Application of the tests described in Section 2 depends on ability to calculate $F(X_1)$ and $F(X_r)$ (Y_1 and Y_r). This means that the population distribution of the measured character must be known "exactly." This is, of course, never the case in practice. However, we note that (a) a sufficiently accurate approximation to the population CDF will serve quite well, and (b) this approximation needs to be good only for the ranges of values likely to be taken most frequently by X_1 and X_r. It is of interest to ascertain the likely effects of using supposedly known (but actually inaccurate) values for F(x) in applying the tests described in Section 2. The results of some enquiries on these lines were presented in [7] and [8]; a synthesis of this work is contained in [9].

Although numerical calculations were confined to cases of normal and exponential CDF's, some general formulae for location-scale parameter families of distributions were obtained. Because of their relative generality we reproduce them here.

Denote the population CDF by

$$F(x) = g\left(\frac{x-\xi}{\theta}\right) \qquad (\theta > 0) \tag{12}$$

and suppose that the CDF used in calculating the tests is

$$F^*(x) = g\left(\frac{x-\xi^*}{\theta^*}\right) \qquad (\theta^* > 0) \tag{13}$$

--that is the correct form of CDF, but possibly incorrect values ξ^*, θ^* for the location and scale parameters ξ, θ respectively.

Then the statistics $Y_i^* = F^*(X_i)$ used in the tests will be monotonic increasing functions of $Y_i = F(X_i)$.

In fact

$$\begin{aligned} Y_i^* &= g\left(\frac{X_i-\xi^*}{\theta^*}\right) \\ &= g\left(\frac{\theta}{\theta^*}\left[\frac{X_i-\xi}{\theta} - \frac{\xi^*-\xi}{\theta}\right]\right) \\ &= g\left(\frac{\theta}{\theta^*} g^{-1}(Y_i) - \frac{\xi^*-\xi}{\theta^*}\right) \\ &= \delta\left(Y_i; \frac{\theta}{\theta^*}, -\frac{\xi^*-\xi}{\theta^*}\right) \end{aligned} \tag{14}$$

where $\delta(z;A,B) = g(Ag^{-1}(z)+B)$.

Inverting (14) we find

$$Y_i = \delta\left(Y_i^*; \frac{\theta^*}{\theta}; \frac{\xi^*-\xi}{\theta^*}\right) . \tag{15}$$

(Generally

$$\delta^{-1}(z;A,B) = \delta(z;A^{-1},-BA^{-1}).) \tag{16}$$

Tests for Censoring from Above or Below

The inequality $Y_1^* > K_1$ is equivalent to $Y_1 > K_1^*$ where

$$K_1^* = \delta\left(K_1; \frac{\theta}{\theta^*}, -\frac{\xi^*-\xi}{\theta^*}\right) . \tag{17}$$

The effect of using an incorrect CDF is therefore to produce a test with a different significance level

$$\alpha^* = \Pr[Y_1 > K_1^* | H_{0,0}] \tag{18}$$

but which is the optimal test at this level. This last result applies generally (not only to location-scale parameter families), and also for censoring from above.

Test for Symmetrical Censoring

The probability of rejection using test (10) is

$$E[I_{1-\gamma_2(Y_r)}(r-1, s_0+1)] \tag{19}$$

where $I_p(a,b) = [B(a,b)]^{-1} \int_0^p t^{a-1}(1-t)^{b-1}dt$ is the incomplete beta function,

$$\gamma_2(y) = \begin{cases} y^{-1}\delta\left(K_2\{1-\delta(y; \frac{\theta}{\theta}, -\frac{\xi^*-\xi}{\theta^*})\}; \frac{\theta^*}{\theta}, \frac{\xi^*-\xi}{\theta}\right) \\ \quad \text{for } y \text{ in the interval } \delta(1 \pm \sqrt{1-4K_2}; \frac{\theta^*}{\theta}, \frac{\xi^*-\xi}{\theta}) \\ 1 \quad \text{otherwise} \end{cases} \tag{20}$$

and, of course, Y_r has a standard beta distribution with parameters $r+s_0$, s_r+1.

General Purpose Test

The probability of rejection using test (11) is

$$E[I_{1-\gamma_3(Y_r)}(r-1, s_0+1)] \tag{21}$$

where

$$\gamma_3(y) = \begin{cases} y^{-1}\delta\left(K_3-1+\delta(y; \frac{\theta}{\theta^*}, -\frac{\xi^*-\xi}{\theta^*}); \frac{\theta^*}{\theta}, \frac{\xi^*-\xi}{\theta}\right) \\ \quad \text{for } y > \delta(1-K_3; \frac{\theta^*}{\theta}, \frac{\xi^*-\xi}{\theta}) \\ 0 \quad \text{otherwise.} \end{cases} \tag{22}$$

The significance level of this test $(s_0 = s_r = 0)$ is thus

$$[\delta(1-K_3; \frac{\theta^*}{\theta}, \frac{\xi^*-\xi}{\theta})]^r + r\int_\eta^1 \{y-\delta(K_3-1+\delta(y; \frac{\theta}{\theta^*}, -\frac{\xi^*-\xi}{\theta^*}); \frac{\theta^*}{\theta}, \frac{\xi^*-\xi}{\theta})\}^{r-1} dy \tag{23}$$

where $\eta = \delta(1-K_3; \frac{\theta^*}{\theta}, \frac{\xi^*-\xi}{\theta})$.

Some numerical results based on these formulae, for normal population distributions are shown in Tables 2.

These formulae (for power) can be averaged over appropriate joint distributions of ξ^* and θ^*, where these latter are estimators of ξ and θ based on various kinds of available information - for example, a (complete) random sample of specified size.

r	$\frac{\theta^*}{\theta} \backslash \frac{\lvert\xi^*-\xi\rvert}{\theta}$ = 0		0.05	0.10	0.25	0	0.05	0.10	0.25	
	$s_0=0$, $s_r=0$ (Significance level)					$s_0=1$, $s_r=1$				
r=10	1.1	.090	.090	.089	.084	.404	.404	.401	.385	
	1	.050	.050	.049	.046	.276	.275	.273	.258	
	0.9	.025	.025	.024	.022	.167	.166	.165	.153	
r=20	1.1	.116	.115	.114	.109	.502	.501	.499	.483	
	1	.050	.050	.049	.046	.306	.305	.303	.287	
	0.9	.019	.017	.017	.015	.146	.146	.144	.134	
	$s_0=3$, $s_r=3$					$s_0=2$, $s_r=4$				
r=10	1.1	.901	.901	.899	.887	.869	.862	.854	.820	(+)
							.874	.878	.883	(-)
	1	.807	.806	.803	.783	.761	.751	.739	.691	(+)
							.769	.775	.781	(-)
	0.9	.665	.663	.659	.631	.610	.598	.583	.525	(+)
							.620	.627	.632	(-)
r=20	1.1	.959	.959	.958	.952	.941	.938	.934	.917	(+)
							.944	.946	.948	(-)
	1	.871	.870	.869	.854	.837	.830	.822	.788	(+)
							.842	.846	.848	(-)
	0.9	.690	.689	.685	.660	.642	.633	.621	.571	(+)
							.650	.655	.655	(-)

(+) → $\xi^* > \xi$

(-) → $\xi^* < \xi$

TABLE 2.1 Significance Level and Power of Test for Symmetrical Censoring (10) (Nominal Significance Level 5%) Normal Population: Mean ξ, Standard Deviation θ.

r	$\frac{\theta^*}{\theta} \backslash \frac{\lvert\xi^*-\xi\rvert}{\theta} =$	0	0.05	0.1	0.25	0	0.05	0.1	0.25
	$s_0+s_r = 0$ (Significance level)					$s_0+s_r = 2$			
r=10	1.1	.080	.080	.082	.098	.325	.344	.369	.461
	1	.050	.051	.052	.066	.237	.261	.281	.372
	0.9	.030	.031	.033	.044	.166	.190	.213	.299
r=20	1.1	.105	.106	.111	.139	.423	.452	.491	.606
	1	.050	.051	.053	.072	.267	.300	.323	.438
	0.9	.021	.021	.024	.042	.151	.166	.203	.336
	$s_0+s_r = 4$					$s_0+s_r = 6$			
r=10	1.1	.605	.634	.669	.775	.809	.833	.860	.928
	1	.502	.546	.577	.698	.730	.774	.799	.887
	0.9	.451	.508	.534	.659	.644	.698	.740	.845
r=20	1.1	.735	.769	.812	.897	.910	.929	.951	.983
	1	.577	.633	.662	.787	.814	.860	.877	.945
	0.9	.418	.446	.517	.703	.693	.718	.781	.908

TABLE 2.2 Significance Level and Power of General Purpose Test for Censoring (11) (Nominal Significance Level 5%) Normal Population: Mean ξ, Standard Deviation θ.

In the case of exponential populations with DCF

$$F(x) = 1 - \exp(-x/\theta) \quad (x>0;\ \theta>0)$$

with θ estimated by the arithmetic mean, θ^*, of an independent complete random sample of size N, (so that θ^* is distributed as $(2N)^{-1}\theta\cdot(\chi^2$ with 2N degrees of freedom)), numerical values of the power have been evaluated, and presented in [9]. Some values are given, in Table 3, for test (9.1) with the value K_1 equal to $1 - \exp\{-r^{-1}N(\alpha^{-1/N}-1)\}$ (*not* $1-\alpha^{1/r}$) with $\alpha = 0.05$. This gives an *actual* significance level equal to 0.05.

4. TESTS BASED ON PARTIAL KNOWLEDGE OF POPULATION DISTRIBUTION

The need to have "exact" knowledge of the population distribution of the measured character is an unfortunate feature of the tests described in Section 2. Even though it is possible to make some allowance for possible inaccuracy in CDF, it is natural to enquire whether useful tests can be based on only *partial* knowledge of the population distribution. As pointed out in Section 1, *some* knowledge is essential. Our question is - how much is needed?

r =		5				10			
N =		5	10	25	50	5	10	25	50
s_0	s_r								
2	0	.2408	.2736	.2973	.3059	.2731	.3132	.3420	.3526
1	1	.0964	.0993	.1008	.1013	.1241	.1327	.1385	.1405
4	0	.4396	.5188	.5764	.5975	.5174	.6077	.6701	.6923
3	1	.2678	.3075	.3370	.3480	.3537	.4133	.4371	.4733
6	0	.5965	.7027	.7746	.7996	.7008	.8054	.8678	.8877
5	1	.4368	.5200	.5826	.6062	.5682	.6689	.7376	.7617
4	2	.2874	.3329	.3677	.3810	.4212	.4981	.5548	.5759
8	0	.7099	.8216	.8885	.9096	.8196	.9107	.9542	.9659
7	1	.5751	.6855	.7634	.7913	.7253	.8325	.8946	.9137
6	2	.4346	.5210	.5881	.6139	.6090	.7173	.7898	.8147
r =		15				20			
N =		5	10	25	50	5	10	25	50
s_0	s_r								
2	0	.2858	.3286	.3594	.3706	.2926	.3369	.3686	.3802
1	1	.1355	.1465	.1541	.1568	.1417	.1541	.1626	.1656
4	0	.5484	.6414	.7042	.7261	.5650	.6591	.7217	.7434
3	1	.3895	.4562	.5043	.5219	.4091	.4791	.5292	.5474
6	0	.7402	.8402	.8965	.9137	.7608	.8574	.9100	.9257
5	1	.6204	.7229	.7893	.8118	.6481	.7503	.8145	.8358
4	2	.4794	.5656	.6270	.6492	.5115	.6015	.6641	.6864
8	0	.8575	.9362	.9700	.9784	.8763	.9477	.9766	.9834
7	1	.7798	.8776	.9289	.9437	.8073	.8985	.9436	.9561
6	2	.6782	.7851	.8509	.8723	.7144	.8180	.8787	.8977

TABLE 3 Power of Test (9.2) (for censoring from below) with Respect to H_{s_0,s_r}. Exponential Population: θ Estimated by $\theta^* \sim (2N)^{-1}\theta\chi^2_{2N}$.

It is also natural to suppose this partial knowledge might take the form of a random sample, either uncensored, or censored *in a known way*, from the population. If the sample is large and uncensored (or only slightly censored) we approach the situation of "exact" knowledge of the population distribution.

In [12] and [13] it is supposed that the "partial knowledge" is not of so definite a character. In the simplest case, it is supposed that one of two sets of observations represents a complete random sample, while the other may or may not be censored (in one of the ways we have described).

If it is known that the sample S_1 is complete and there are L_1 (G_1) members of S_1 which are less (greater) than any members of S_2, the likelihood function, under H_{s_0,s_r}, is

$$L(s_0,s_r) = \binom{L_1+s_0}{s_0}\binom{G_1+s_r}{s_r}\binom{r_1+r_2-L_1-G_1-2}{r_2-2}\Big/\binom{r_1+r_2+s_0+s_r}{r_1} . \tag{24}$$

For testing $H_{0,0}$ (no censoring) against H_{s_0,s_r} the likelihood ratio is

$$\frac{L(s_0,s_r)}{L(0,0)} = \frac{(L_1+1)^{[s_0]}(G_1+1)^{[s_r]}}{(r_1+r_2+1)^{[s_0+s_r]}} \frac{(r_2+1)^{[s_0+s_r]}}{s_0!s_r!} . \tag{25}$$

Hence we use a test of $H_{0,0}$ with a critical region

$$(L_1+1)^{[s_0]}(G_1+1)^{[s_r]} > K \tag{26}$$

(since the likelihood ratio is proportional to the left hand side of (26).)

For censoring from above (or below) this is equivalent to

$$G_1 \text{ (or } L_1) > K_1 . \tag{27}$$

For symmetrical censoring we obtain

$$(L_1+1)^{[s]}(G_1+1)^{[s]} > K_2 \tag{28}$$

which depends on s, in contradistinction to (10). However, the test is not greatly dissimilar from one using the critical region

$$(L_1+1)(G_1+1) > K_3 \tag{29a}$$

or even

$$L_1G_1 > K_4 . \tag{29b}$$

For a general purpose test, it is natural to consider the critical region

$$L_1+G_1 > K_5 . \tag{30}$$

Since L_1, G_1 are discrete random variables, it is not usually possible to choose the K's to give prespecified significance levels.

If r_2 is kept fixed, while r_1 (the size of the known complete random sample) tends to infinity then

$$r_1L_1 \sim F(X_1);\ r_1G_1 \sim 1-F(X_{r_2})$$

X_1 and X_{r_2} being the least and greatest values in S_2, respectively. The tests (28), (29) and (30) tend to those appropriate when $F(\cdot)$ is known, as is to be expected.

We will give some details of calculation of power for the test (30). From (24)

$$\Pr[L_1+G_1=t|s_0,s_r] = \binom{r_1+r_2+s_0+s_r}{r_1}^{-1}\binom{r_1+r_2-t-2}{r_2-2}\sum_{g=0}^{t}\binom{t-g+s_0}{s_0}\binom{g+s_r}{s_r}$$

$$= \binom{r_1+r_2+s_2+s_r}{r_1}^{-1}\binom{r_1+r_2-t-2}{r_2-2}\binom{t+s_0+s_r+1}{s_0+s_r+1} . \tag{31}$$

This will be recognized as the distribution of number of exceedances, as given by Sarkadi ([14] eq. (1) with $n_1 = r_2+s$, $n_2 = r_1$, $m = s+2$). We have to choose K_5 so that $\sum_{t=K_5}^{r_1} \Pr[L_1+G_1=t|0,0]$ is as near the required significance level as possible. The power with respect to H_{s_0,s_r} is then

$$\sum_{t=K_5}^{r_1} \Pr[L_1+G_1=t|s_0,s_r] .$$

As with (10), it depends only on the sum (s_0+s_r), not on s_0,s_r separately. Some numerical results for $r_2 = 5(5)15$ with various sizes (r_1=10,20,50,100) of the complete sample are shown in Table 4. It should be noted that greater power is to be expected using (27) or (28) in the cases when censoring is indeed from above (or below) and symmetrical, respectively.

r_2	r_1	$s_0+s_r =$ K_5	0 Significance level	2 Power	4 Power	6 Power
5	10	8	.0470	.1448	.2585	.3673
	20	15	.0403	.1392	.2663	.3942
	50	34	.0532	.1872	.3566	.5191
	100	67	.0510	.1857	.3613	.5320
10	10	5	.0704	.2214	.3888	.5359
		6	.0286	.1099	.2235	.3441
	20	9	.0621	.2271	.4274	.6037
		10	.0369	.1535	.3187	.4854
	50	21	.0538	.2278	.4587	.6624
	100	41	.0503	.2270	.4714	.6873
15	10	4	.0640	.2036	.3638	.5096
		5	.0225	.0910	.1931	.3077
	20	7	.0548	.2092	.4059	.5853
	50	16	.0448	.2081	.4413	.6547
	100	30	.0472	.2299	.4913	.7177

TABLE 4. Power of Test (30). Approximate Significance Level 0.05.

A useful relation in the calculations is

$$\Pr[L_1+G_1=t+1|s_0,s_r] = \Pr[L_1+G_1=t|s_0,s_r] \cdot \frac{(r_1-t)(s_0+s_r+2+t)}{(r_1+r_2-t-2)(1+t)} \qquad (32)$$

Even when it is only known that just one of the two samples (S_1,S_2) is censored (but not which one, if any) a useful distribution-free procedure can be constructed to decide which is the censored sample. This is a main topic of [12] and [13]. We restrict ourselves to the relatively simple case $r_1 = r_2 = r$, say - equal sizes of the two observed sets, and define $G_i(L_i)$ to be the number of values from S_i which are greater (less) than any values from S_{3-i} (i=1,2). Clearly one of G_1 and G_2 (L_1 and L_2) must be zero and the other must not be zero. The likelihood ratio approach leads us to the following tests. (H_i denotes the hypothesis that S_i is the censored sample.)

Censoring from Above (Below) ([13])

Accept H_i if G_{3-i} $(L_{3-i}) > 0$. (33)

This is, in fact, *identical* with the procedure which would be used if the population CDF were known. The probability of correct decision is

$$1 - \frac{(s+r)!(2r-1)!}{(r-1)!(s+2r)!}$$

where s is the number of observations removed.

If we have two *sequences* $S_{i1},S_{i2},S_{i3},\ldots,S_{im}$ (i=1,2) and it is known that (i) in just one of the two sequences each sample $S_{i1},S_{i2},\ldots$ has been censored in the same way from above (below) *and* (ii) the population CDF is the same in S_{1j} and S_{2j} but *may change with* j, then the best decision procedure is

$$\text{Accept } H_1 \ (H_2) \text{ if } \prod_{j=1}^{m} \left[\frac{(G_{2j}+1)^{[s]}}{(G_{1j}+1)^{[s]}}\right] > (<) \ 1 . \qquad (34)$$

Although this procedure depends on s, it will not differ greatly (for small m) from the procedure based on s = 1. For small m, probabilities of correct decision, and no decision (when the ratio in (26) equals 1) are easily evaluated. For example when m = 2:

$$\text{Probability of correct decision} = \left\{1 - \frac{(s+r)!(2r-1)!}{(s+2r)!(r-1)!}\right\}^2$$

$$+ \ 2\binom{2r+s}{r}^{-1} \sum_{j=2}^{r} \binom{s+j}{s} \binom{2r-j-1}{r-1} \sum_{h=1}^{j-1} \binom{2r-h-1}{r-1} . \qquad (35)$$

$$\text{Probability of no decision} = 2\binom{2r+s}{s}^{-2} \sum_{j=1}^{r} \binom{s+j}{s} \binom{2r-j-1}{r-1}^2. \quad (36)$$

Some numerical values are given in Tables 5.1 and 5.2. It will be noticed that the probability of correct decision is less when m = 2 than when m = 1, for s = 1, but so is the probability of *incorrect* decision. ("No decision" cannot occur when m > 1.)

r\s =	1	2	3	4	5
5	0.727	0.841	0.902	0.937	0.958
10	0.738	0.857	0.919	0.953	0.972
15	0.742	0.863	0.925	0.958	0.976
20	0.744	0.866	0.928	0.961	0.978
∞	0.750	0.875	0.9375	0.969	0.984

TABLE 5.1 Detection of Which of Two Sequences is Censored from Above (or from Below). Single Pair of Samples (m=1). Probabilities of Correct Decision.

r\s =	1		2		3		4		5	
	Correct	No	Correct	No	Correct	No	Correct	No	Correct	No
5	0.692	0.068	0.842	0.039	0.915	0.023	0.953	0.013	0.972	0.008
10	0.717	0.058	0.870	0.030	0.940	0.015	0.972	0.007	0.986	0.004
15	0.726	0.055	0.880	0.027	0.948	0.012	0.977	0.005	0.990	0.003

TABLE 5.2 Detection of Which of Two Sequences is Censored from Above (or from Below). Two pairs of Samples (m=2). Probabilities of Correct and No Decision.

If it can be assumed that the common CDF for S_{1j} and S_{2j} does not depend on j, there is, presumably, a discrimination procedure giving even greater chances of correct decision than those shown in Tables 5.

Symmetrical Censoring ([12])

For a single pair of samples, the best procedure is

$$\text{"Accept } H_1 \ (H_2) \text{ if } \frac{(L_2+1)^{[s]}(G_2+1)^{[s]}}{(L_1+1)^{[s]}(G_1+1)^{[s]}} > (<) \ 1." \quad (37)$$

For m pairs of samples, it is

$$\text{"Accept } H_1 \ (H_2) \text{ if } \prod_{j=1}^{m} \left[\frac{(L_{2j}+1)^{[s]}(G_{2j}+1)^{[s]}}{(L_{1j}+1)^{[s]}(G_{1j}+1)^{[s]}}\right] > (<) \ 1 \ ." \quad (38)$$

As with (34), procedure (38) does not depend greatly on s, at least for small m. For m = 1, the rule can be expressed

"Accept H_1 (H_2) if L_2+G_2 > (<) L_1+G_1." (39)

which does not depend on s at all.

Tables 6.1 and 6.2 show probabilities of correct, no and incorrect decisions with m - 1 and m = 2 respectively. There is a marked increase in the probability of correct decision when m increases from 1 to 2.

	s=1		s=2	
r	Prob. correct	Prob. no decision	Prob. correct	Prob. no decision
5	0.619	0.194	0.742	0.152
10	0.690	0.129	0.835	0.076
15	0.709	0.117	0.858	0.063
20	0.717	0.112	0.869	0.058
30	0.726	0.107	0.879	0.052
∞*	0.743	0.097	0.899	0.043

*Values for r = ∞ were obtained by harmonic extrapolation.

TABLE 6.1 Properties of Distribution-Free Test for Detection of Samples Censored Symmetrically. Single pair of samples (m = 1).

	s=1		s=2	
r	Prob. correct	Prob. no decision	Prob. correct	Prob. no decision
4	0.791	0.117	0.897	0.076
5	0.832	0.084	0.936	0.044
6	0.851	0.066	0.954	0.029

TABLE 6.2 Properties of Distribution-Free Test for Detection of Samples Censored Symmetrically. Two pairs of samples (m = 2).

General Purpose Discrimination ([13])

We now suppose that s_0 and s_r are not known (though are supposed to remain the same throughout). For a single pair of samples natural analogy to test (11) would be to use the rule

"Accept H_1 (H_2) if G_1+L_1 < (>) G_2+L_2." (40)

This is, in fact, the same as rule (39). If m is greater than 1, we might combine the values of

$$D_{ij} = G_{ij} + L_{ij} \quad (j=1,2,\ldots,m)$$

either by multiplication or summation, leading to rules

$$\text{"Accept } H_1 \ (H_2) \text{ if } \prod_{j=1}^{m} (D_{1j}+1) < (>) \prod_{j=1}^{m} (D_{2j}+1)\text{"} \tag{41}$$

or

$$\text{"Accept } H_1 \ (H_2) \text{ if } \sum_{j=1}^{m} D_{1j} < (>) \sum_{j=1}^{m} D_{2j}\text{."} \tag{42}$$

An approximate assessment of the properties of procedure (42) can be based on normal approximation to the distribution of

$$T_m = \sum_{j=1}^{m} (D_{1j} - D_{2j}) .$$

The calculation of the first and second moments of T_m is somewhat involved. The results are

$$m^{-1}E[T_m] = \binom{2r+s_0+s_r}{r}^{-1} [\sum{}'\{\binom{2r+s}{r+1} - \binom{r+s}{r+1}\} - (s_0+s_r+2)\binom{2r+s_0+s_r}{r-1}]$$

$$m^{-1}\text{var}[T_m] = \binom{2r+s_0+s_r}{r}^{-1} [\sum{}'\{2\binom{2r+s}{r+2} + \binom{2r+s}{r+1} - 2(s+1)\binom{2r+s}{s} - 2\binom{r+s}{r+2}$$

$$- (2r+1)\binom{r+s}{r+1} + 2(s+1)\binom{r+s}{r}\}$$

$$+ (s_0+s_r+2)\binom{2r+s_0+s_r}{r-1} + (s_0+s_r+2)(s_0+s_r+3)\binom{2r+s_0+s_r}{r-2}$$

$$+ 2\binom{2r}{r-2}] - [m^{-1}E[T_m]]^2 \tag{44}$$

where $\sum{}'$ means summation over $s = s_0$, $s = s_r$.

From the normal approximation one can estimate how big m needs to be to ensure a desired probability of correct choice between H_1 and H_2. Results of such calculations are shown in Table 7.

It should again be noted that the tests discussed in this Section apply even if the common (unknown) population of the two samples in a pair changes from one pair to another. Presumably if it could be assumed that the distribution is the same for all the samples, a more efficient procedure could be used, and the probabilities obtained for our procedures improved upon.

Another related question is to test the hypothesis that one of two samples *is* censored, it being known that no more than one, if any, may be censored. Work is at present in progress on analogous problems for more than two sequences of samples, with various patterns of censoring and various limitations on numbers of sequences subject to censoring.

r	s'	s''	P=0.99	P=0.999
5	1	1	15	26
	2	0	21	37
	1	0	28	50
	0	0	-	-
10	2	1	6	11
	3	0	8	14
	1	1	11	20
	2	0	14	25
	1	0	41	72
	0	0	-	-
15	2	2	4	6
	3	1	4	7
	4	0	5	9
	2	1	6	10
	3	0	7	12
	1	1	11	18
	2	0	13	22
	1	0	38	66
	0	0	-	-
∞	2	2	3	6
	3	1	4	6
	4	0	4	7
	2	1	5	9
	3	0	6	10
	1	1	9	16
	2	0	11	18
	1	0	32	57
	0	0	-	-

TABLE 7 Approximate Number of Pairs of Samples Needed to Ensure Minimum Probability, P, of Correct Decision Using (42).

5. MULTIVARIATE PROBLEMS

There is no lack of other problems for current and future study in the field of multivariate censoring. Some of these problems are described in [11]. Simplest are those in which censoring, if present, is in a known variable. These can be attacked by the methods appropriate to univariate data, using, of course, only the values of the variable determining the censoring.

A new type of problem arises if it is known that censoring occurs in some one (or a known number) of the variables, but it is not which variable(s) is(are) subject to censoring. Determination of the censored set is an interesting problem. A related, somewhat more difficult, problem arises when it is desired to test whether such censoring is, indeed, present. The censoring itself may, of course, be of any of the types we have discussed in this paper, or, indeed, may be more general.

Other problems, which do not have counterparts in univariate data, include (i) trying to establish whether censoring has been based on values of some linear function of the variable values, and (ii) to test $H_{0,0}$ given only values of a variable which is correlated with a variable with respect to which censoring has operated, but values of which are unrecorded.

In the latter case, one might use the same statistics as those used when the recorded values are those of the variable on which censoring is based. If the null hypothesis ($H_{0,0}$ - no censoring) is valid, the statistics would have the same distribution for either variables, since both are based on complete random samples. There would, however, be some loss of power because the ordering of the recorded values will not in general agree with that of the censoring variable.

These examples suffice to indicate the considerable variety of problems associated with detection of censoring in multivariate data. In many cases it will be enough to consider each variable separately and use univariate techniques, together with some consideration of pairs of variables. Initial research, therefore, will be concentrated on bivariate problems.

6. RELATED WORK

This paper is primarily concerned with my own work on the subjects considered. For completeness, I now give a brief summary of some related work. References [15] and [16] relate directly to [1] and [10] respectively.

In [1], estimation of sample size from the r observed values, known to be the least r in a random sample, was considered. In [15], D.G. Hoel constructed a sequential procedure, with r increasing as time progresses, to reach conclusions about ultimate sample size.

In [10], on the other hand, the total sample size (n) is known, but there are only r values (randomly chosen) out of the n available. It is desired to estimate the rank order of any given one of these r among the n in the complete sample. Two unbiased estimators - one distribution-free, the other requiring a knowledge of the population distribution - were suggested, and optimal linear combinations of the two discussed. In [16], Campbell and Hollander obtain optimal estimators (with mean square error as criterion) for these cases.

Blumenthal and Sanathanan [17] have in preparation a book on topics in this field which includes their own work ([18] and [19] are typical references) and also much of the work described in the present paper. (In general their work exploits the specific parametric form of distribution assumed in the population making use, for example, of the existence of sufficient statistics for the parameters).

REFERENCES

[1] N.L. Johnson (1962). Estimation of sample size, *Technometrics, 4,* 59-67.

[2] ___________ (1966). Tests of sample censoring, *Proc. 20th Tech. Conf. Amer. Soc. Qual. Control,* 699-703.

[3] ___________ (1967). Sample censoring, *Proc. 12th Conf. Des. Exp. Army Res.,* 403-424.

[4] ___________ (1970). A general purpose test of censoring of extreme sample values, *S.N. Roy Memorial Volume,* pp. 377-384. Chapel Hill: Univ. N. Carolina Press.

[5] ___________ (1971). Comparison of some tests of sample censoring of extreme values, *Austral. J. Statist., 12,* 1-6.

[6] ___________ (1972). Inferences on sample size: Sequences of samples, *Trab. Estad., 23,* 85-110.

[7] ___________ (1973). Robustness of certain tests of censoring of extreme sample values, UNC Institute of Statistics, Mimeo Series *866*.

[8] ___________ (1974). *Ibid.*, II. Some exact results for exponential populations, *Ibid., 940.*

[9] ___________ (1974). Study of possibly incomplete samples, *Proc. 5th Internat. Conf. Prob. Statist., Brasov (Romania)*, pp. 59-73: Bucharest: Acad. Rep. Soc. Romania.

[10] ___________ (1974). Estimation of rank order, UNC Institute of Statistics, Mimeo Series *931*.

[11] ___________ (1975). Extreme sample censoring problems with multivariate data, I., *Ibid., 1010.*

[12] ___________ (1978). Completeness comparisons among sequences of samples, *H.O. Hartley 65th Birthday Volume*, (pp. 259-275), New York: Academic Press.

[13] ___________ (1977). *Ibid.* II. Censoring from above or below, and general censoring, UNC Institute of Statistics Mimeo Series *1103* (to appear in *J. Statist. Planning and Inference*).

[14] Sarkadi, K. (1960). On the median of the distribution of exceedances, *Ann. Math. Statist., 31,* 225-226.

[15] Hoel, D.G. (1968). Sequential testing of sample size, *Technometrics, 10,* 331-341.

[16] Campbell, G. and Hollander, M. (1978). Rank order estimation with the Dirichlet prior, *Ann. Statist., 6,* 142-153.

[17] Blumenthal, S. and Sanathanan, L. (1979?). *Statistical Methods for Incomplete and Truncated Data.* New York: Wiley.

[18] Blumenthal, S. and Marcus, R. (1975). Estimating population size with exponential failure, *J. Amer. Statist. Assoc.*, *28*, 913-922.

[19] Sanathanan, L.P. (1972). Estimating the size of a multinomial population, *Ann. Math. Statist*, *43*, 142-152.

The author was supported by the Army Office of Research under Contract DAAG29 74 C 0030.

Department of Statistics
University of North Carolina
Chapel Hill, NC 27514

Robust Estimation for Time Series Autoregressions

R. Douglas Martin

1. INTRODUCTION

There now exists a large and rapidly growing statistical literature dealing with robust estimation. A large portion of this literature treats location and linear regression models with independent and identically distributed errors. A relatively small number of notable contributions deal with robust estimation of covariance matrices. Research on robust estimation in the time series context has lagged behind, and perhaps understandably so in view of the increased difficulties imposed by dependency and the considerable diversity in qualitative features of time series data sets. This paper discusses some theory and methodology of robust estimation for time series having two distinctive types of outliers. As a point of departure we begin by mentioning some of the central concepts in robustness that have been developed primarily in the independent observations context. Since it is assumed that most readers will have some familiarity with robust estimation the discussion will be quite brief. Further background material may be found in Hogg's (1978) tutorial article in this volume, and in P. Huber's (1977) monograph.

Robustness Concepts

Loosely speaking, a robust estimate is one whose performance remains quite good when the true distribution of the data deviates from the assumed distribution. Data sets for

ISBN 0-12-438150-2

which one often makes a Gaussian assumption sometimes contain a small fraction of unusually large values or "outliers". More realistic models for such data sets are provided by distributions which are heavy-tailed relative to a nominally Gaussian distribution. Quite small deviations from the Gaussian distribution, in an appropriate metric, can give rise to heavy-tailed distributions and correspondingly potent outliers in the generated samples. Since such deviations from a Gaussian distribution can have serious adverse effects on Gaussian maximum-likelihood procedures, it is not surprising that attention has been focused primarily on robustness over families of distributions which include both the nominal Gaussian model and heavy-tailed distributions. Relatively little attention has been devoted to other nominal parametric models or other alternatives, such as short-tailed distributions and asymmetric distributions (cf., Hampel, 1968, and Hogg, 1977).

The seminal papers by Tukey (1960), Huber (1964), and Hampel (1971) provide more precise notions of robustness, namely <u>efficiency</u> robustness, <u>min-max</u> robustness and <u>qualitative</u> robustness, respectively.

An efficiency robust estimate is one whose (appropriately defined) efficiency is "high" at a variety of strategically chosen distributions. Huber's min-max robust location estimates minimize the maximum asymptotic variance over certain uncountably infinite families of distributions.

Hampel's qualitative robustness is a rather natural equicontinuity requirement. Suppose F_o, F are underlying distributions for i.i.d. observations, $\{T_n\}$ a sequence of estimators indexed by sample size, and $\mathcal{L}(T_n, F_o)$, $\mathcal{L}(T_n, F)$ the distributions or "laws" of T_n under F_o, F respectively. Then $\{T_n\}$ is qualitatively robust at F_o if the sequence of maps $F \to \mathcal{L}(T_n, F)$ is equicontinuous at F_o in an appropriate metric.

Hampel also introduced the <u>influence curve</u> (1968, 1974) and the <u>breakdown point</u>, both useful concepts in robustness. Hogg's (1978) article in this volume provides an introduction to influence curves.

The breakdown point of an estimate is essentially the largest fraction of the data, over all combinations for each fraction, which can be taken to $\pm\infty$ without taking the (asymptotic) value of the estimate to the boundary of the parameter space. In location problems the boundary points are $\pm\infty$. For estimating a correlation coefficient the points ± 1 might appear to be the natural boundary points. However, under the hypothesis that the true correlation is non-zero it would be reasonable to include the origin as well as ± 1 in the boundary set. This possiblity is quite a natural one when estimating autoregressive parameters, as we shall see in Section 6.

Hampel argues persuasively that a highly robust estimate should have the following two properties, among others: (i) the highest possible breakdown point -- 1/2 is effectively an upper bound, and (ii) a bounded and continuous influence curve which "redescends" to zero at finite points.

Since the remainder of the paper is concerned primarily with time series autoregressive models, some comments about robust regression for the usual linear model are appropriate. Let

$$y_i = z_i^T\beta + \varepsilon_i \qquad i = 1,\dots,n \tag{1.1}$$

denote the linear model where $z_i^T = (z_{i1},\dots,z_{iq})$ and $\beta^T = (\beta_1,\dots,\beta_q)$. An M-estimate $\hat{\beta}_M$ of β (Relles, 1968; Huber, 1973) is a solution of the minimization problem

$$\min_{\beta'} \sum_{i=1}^{n} \rho\left(\frac{y_i - z_i^T\beta'}{s}\right) \tag{1.2}$$

where s is a scale parameter for the errors. In practice s must be estimated robustly as well as β. However, to simplify the discussion here let's assume that $s = 1$. With $\psi = \rho'$ differentiation of (1.2) gives the estimating equation

$$\sum_{i=1}^{n} z_i \, \psi(y_i - z_i^T\hat{\beta}_M) = 0 \quad . \tag{1.3}$$

Suppose ψ is monotone and bounded. Then for each fixed β' the function $h(z,y) = z\cdot\psi(y - z^T\beta')$ is bounded in the scalar y, but unbounded in $z^T = (z_1,\dots,z_q)$. Correspondingly, it turns out that the influence curve for $\hat{\beta}_M$ is bounded in y and unbounded in z. This feature would be appropriate if one could be sure that the z portion of the model is correctly specified. However, there is always the possibility of z mis-specification; for example, there might be a small fraction of gross errors in variables. When such deviations from the ideal model are a possibility the unbounded character of the influence curve can have unpleasant consequences. This is particularly relevant for time series autoregressions observed with additive outliers, as will be evident in Section 6.

The possibility of bounding the influence function in z as well as in y has been alluded to by Huber (1973), suggested by Mallows (1973, 1976) and advocated by Hampel (1973, 1975). One possible bounded influence version of (1.4) is the generalized M-estimate (GM-estimate) $\hat{\beta}_{GM}$ obtained by solving

$$\sum_{i=1}^{n} W(z_i)\cdot z_i \cdot \psi(y_i - z_i^T \hat{\beta}_{GM}) = 0 \tag{1.4}$$

where the z_i are robustly centered and $W(z)\cdot z$ is a bounded and continuous function of z. Related recent references are Denby and Larsen (1977), Hill (1977), and Maronna and Yohai (1977). Autoregression versions of (1.4), discussed in Section 6, appear to be quite useful for robust estimation of autoregressive models in the presence of additive outliers.

It is known (Andrews et al., 1972; Denby and Larsen, 1977) that the use of non-monotone "redescending" psi-functions ψ, such as Tukey's bisquare function ψ_{BS} yield higher efficiencies at extremely heavy-tailed distributions than do monotone versions such as Huber's ψ_H (definitions of ψ_{BS} and ψ_H are given in Hogg's (1978) article in this volume as well as in a number of other articles in the robustness literature). Furthermore a redescending psi-function such as ψ_{BS} results in an influence curve which is bounded in z and y. However the estimating equations (1.3) and (1.4) will then have multiple roots which may be quite troublesome.

The iterated weighted least squares (IWLS) method of solving M-estimate equations (Hogg, 1978, this volume; Mosteller and Tukey, 1977) may be used in the following reasonable strategy for using a redescending ψ. First use least squares as a starting point for an approximate IWLS solution of (1.3) or (1.4) with a monotone ψ. Then use that solution as a starting point for one or at most two iterations of IWLS based on a redescending ψ. Of course the scale parameter s, which has been suppressed for convenience, will have to be estimated along with β (see Huber, 1973, 1977).

Robustness Concepts for Time Series

The robustness concepts just described have been developed and used primarily in the context of independent and identically distributed observations (possibly vector-valued as when estimating covariance matrices).

For time series parameter estimation problems, efficiency robustness and min-max robustness are concepts directly applicable. Influence curves for parameter estimates may also be defined without special difficulties. Somewhat greater care is needed in defining breakdown points as the detailed nature of the failure mechanism may be quite important. For instance i.i.d. gross errors and highly correlated or patchy gross errors may yield different breakdown points (see Section 6).

A major problem which remains is that of providing an appropriate and workable definition of qualitative robustness in the time series context. Some results in this direction have been provided by Kazakos and Gray (1977). A simple alternative to applying their interesting, but not entirely satisfactory theory is to evaluate estimators via a purely asymptotic version of qualitative robustness. Thus we just drop the equicontinuity part of Hampel's definition and replace estimator finite sample distributions with asymptotic distributions. This is the special sense in which qualitative robustness will be used throughout the remainder of this paper.

Outline of Paper

In the next section we discuss models for time series with outliers, focusing on two particular cases, innovations outliers (IO) and additive outliers (AO). Section 3 discusses the behavior of the usual least-squares estimates of autoregressive parameters for IO models while Section 4 does similarly for M-estimates. Section 5 is concerned with robust estimates of location for IO models. The difficulties posed by AO models for both least-squares and M-estimates is noted in Section 6, and a class of robust generalized M-estimates having bounded influence curves is introduced. In Section 7 we discuss exact and approximate conditional-mean M-estimates. Section 8 is concerned with determining outlier type.

2. TIME SERIES OUTLIER MODELS

Most robustness studies to date have assumed that outliers appear in the i.i.d. observation errors of a location or regression model, and so the outlier generation is conveniently modeled by a symmetric heavy-tailed univariate distribution. Notable exceptions occur in work on robust estimation of covariance matrices (Devlin, Gnanadesikan and Kettenring, 1975; Maronna, 1976) where heavy-tailed multivariate distributions are considered. In these cases the vector observations are still assumed to be independent.

For time series the situation is more complicated since the desire for a complete probabilistic description of either a nearly-Gaussian process with outliers, or the corresponding asymptotic distribution of parameter estimates, will often dictate that one specify more than a single finite-dimensional distribution of the process. It is only in special circumstances that the asymptotic distribution of the estimate will depend only upon a single univariate distribution (e.g., as in Section 4 or 5) or a single multivariate distribution. More typically the asymptotic distribution will depend upon at least a countably infinite set of finite-dimensional distributions. A pertinent example may be found in Martin and Jong (1976).

In the face of such complications it seems imperative to begin by specifying simple outlier generating models. They should be in some sense "nearly" Gaussian and produce sample paths representative of what occurs in practice. This is a task made particularly difficult by the wide variety of time series data types.

Nonetheless there seem to be some distinct outlier types in time series. One possibility is that of isolated or gross-error outliers, which might be due to various recording errors, and for which an i.i.d. gross-error structure may be appropriate. A second possibility is that of patchy type outliers whose behavior appears somewhat or totally unrelated to the behavior of the remainder of the sample. This type might be due to brief malfunctioning of a recording instrument, an inherent behavior of the process, or perhaps other unaccountable affects. A third possibility is that of patchy behavior in which the character of the outliers is consistent with the remainder of the sample path, except perhaps for an initial jump. The latter type probably occur relatively infrequently.

While admitting the limitations of the above simple categorization it seems quite desirable to capture some of the essence of the above kinds of behavior with appropriate formal models. For the first two kinds of behavior the following additive outliers (AO) model may be the simplest appropriate representation. Let x_i be a stationary and purely nondeterministic zero-mean Gaussian random process, μ a location parameter, v_i a process independent of x_i, and whose marginal distribution satisfies $P(v_i=0) = 1-\delta$. Then the observed process is

$$y_i = \mu + x_i + v_i \ , \qquad i = 1,\ldots,n \ . \tag{2.1}$$

An i.i.d. structure for the v_i provides for modeling the first gross error situation, and various patchy v_i processes with correlation structure unrelated to that of x_i will yield a version of the second type of behavior.

The third kind of behavior might be obtained with $v_i \equiv 0$ and the Gaussian assumption on x_i dropped. Since x_i is purely non-deterministic it has the representation

$$x_i = \sum_{\ell=0}^{\infty} h_\ell \, \varepsilon_{i-\ell} \tag{2.2}$$

with ε_i the uncorrelated zero mean innovations sequence and $\sum_{\ell=1}^{\infty} h_\ell^2 < \infty$, assuming that x_i has finite variance (Wold, 1953). Stationary infinite variance processes can also be defined by the right-hand side of (2.2) under appropriate assumptions (Kanter, 1972; Yohai and Maronna, 1977). Now if the ε_i have a heavy-tailed and symmetric distribution we get a special kind of patchy behavior, and in this case the model

$$y_i = \mu + x_i \tag{2.3}$$

is called an innovations outlier (IO) model. If x_i is Gaussian then outliers do not occur, but for convenience (2.3) will still be referred to as a (Gaussian) IO model.

For the special case of pth-order autoregressive x_i

$$y_i = \mu[1-\sum_{\ell=1}^{p} \phi_\ell] + \sum_{\ell=1}^{p} \phi_\ell \, y_{i-\ell} + \varepsilon_i \tag{2.4}$$

and

$$(y_i-\mu) = \sum_{\ell=1}^{p} \phi_\ell (y_{i-\ell}-\mu) + \varepsilon_i \tag{2.4'}$$

are equivalent versions of (2.3).

If the Gaussian y_i process (2.3), parameterized by $\{h_\ell\}$, μ and σ_ε^2 is the nominal parametric model, then the AO and IO models provide distinctive outlier generating non-Gaussian alternatives. Special autoregression cases of these were introduced by Fox (1972). Gastwirth and Rubin (1975) use a special double-exponential first order version of IO model as well as some more complex models. Also, Abraham and Box (1977) use both IO and AO models in a Bayesian context.

It should be noted that when v_i has a mixture distribution (e.g., a contaminated normal distribution) the same is true of y_i in the AO situation. However the mixture distribution for y_i is a rather special kind since it is obtained via convolution with a Gaussian distribution. Use of the usual kind of mixture distribution for y_i would imply the replacement of x_i by v_i rather than adding x_i to v_i. This seems to me to be for the most part less realistic than the AO model.

Obviously more complex models than those given above will be more appropriate for many time series occurring in practice. However, things are complicated enough with just the IO and AO possibilities and I shall concentrate on the autoregression version of them for the remainder of the paper. The treatment of moving-average and autoregressive-moving average models, as well as the treatment of trends and seasonality, are omissions which will have to be accounted for elsewhere.

3. LEAST SQUARES AUTOREGRESSION

It is well known that the least squares estimates of autoregressive parameters $\phi^T = (\phi_1,\ldots,\phi_p)$ have a striking and extreme form of qualitative robustness for finite variance IO models, namely their asymptotic variance-covariance matrix $V_{\phi,LS}$ is independent of the innovations distribution and depends only upon these autoregressive parameters. This fact is somewhat obscured by the common practice of writing

$$V_{\phi,LS} = \sigma_\varepsilon^2 C^{-1} \tag{3.1}$$

where $C_{ij} = \text{COV}\{\tilde{y}_i,\tilde{y}_i\}$, $1 \leq i, j \leq p$ (cf. Mann and Wald, 1943). However, $C = \sigma_\varepsilon^2 \tilde{C}$ where $\tilde{C}$ is the covariance matrix for a unit variance ε_k. Thus $V_{\phi,LS}$ is better written as

$$V_{\phi,LS} = \tilde{C}^{-1} \tag{3.2}$$

where $\tilde{C}$ depends only upon ϕ.

This behavior of the least squares estimates was pointed out by P. Whittle in his important contributions (1953, 1962). In the 1962 paper he introduced the word "robust" in the time series context. The actual definition given by Whittle was that an estimate is robust if its asymptotic distribution is independent of cumulants of order three or higher. This is a stringent definition of robustness relative to the current standard of qualitative robustness introduced by Hampel (1968, 1971).

Whittle also pointed out the Gaussian case optimality of least squares, but hastened to add that this needn't be the case in any specific non-Gaussian situation. However, in the second paper Whittle established an asymptotic max-min property of the least squares estimate (which also has the min-max robustness property) and the least-favorable character of the Gaussian distribution.

On the other hand the least-squares estimate of ϕ provides a striking example of an asymptotically robust equalizer rule which is not efficiency robust. This is easy to see by computing asymptotic efficiencies. A straightforward calculation (Martin, 1978a) shows that the large sample information matrix for ϕ is

$$I_\phi = \tilde{C} \cdot \sigma_\varepsilon^2 \cdot i(g) \tag{3.3}$$

where i(g) is the Fisher information (for location) for a finite variance innovations density g. Taking the p-th root of the ratio of determinants as a multivariate measure of efficiency (cf. Anderson, 1971) gives

$$\mathrm{EFF}(\mathrm{LS},g) = \left(\frac{\det I_\phi^{-1}}{\det V_{\phi,\mathrm{LS}}} \right)^{\frac{1}{p}} = (\sigma_\varepsilon^2 \cdot i(g))^{-1} . \tag{3.4}$$

But this is just the pth-power of the asymptotic efficiency of the sample mean, and the latter is notoriously lacking in efficiency robustness toward heavy-tailed g.

Consideration of the Cramer-Rao lower bound for the first order case

$$V_{CR} = I_\phi^{-1} = \frac{1-\phi^2}{\sigma_\varepsilon^2} \cdot \frac{1}{i(g)} \tag{3.5}$$

makes it quite transparent how heavy-tailed distributions can be quite helpful. For σ_ε^2 can become arbitrarily large in arbitrarily small neighborhoods of the Gaussian distribution while i(g) remains relatively stable. This allows for the possibility of arbitrarily large increases in precision since (under regularity conditions) the bound is attained asymptotically by the maximum-likelihood estimate. This in turn suggests that one might expect to obtain nearly the same increases in precision using an M-estimate for autoregression. The behavior of such estimates is discussed in the next section.

Note the distinction between the behavior of the least squares estimates for location (with i.i.d. errors) and autoregression. In the former case one may suffer losses in the form of inflated estimator variances when the distribution is heavy-tailed. Thus it is imperative in practice to pay the small loss of efficiency insurance premiums associated with good robust estimates at the Gaussian situation in order to obtain high efficiencies at heavy-tailed distribution. In the latter case the decision to pay a premium is more optional since it would not be made for protection against inflated estimator variance but for possible increased precision in the event of heavy tails.

In sharp contrast to this special situation with regard to estimating ϕ, the least squares estimates of location μ and innovations scale s_ε are not robust, the latter having been pointed out by Whittle (1962), and more recently reiterated by Dzhaparidze and Yaglom (1973). The more important problem is often that of estimating μ, and we return to it in Section 5.

The above discussion has been for finite variance innovations. It should be remarked that appropriately defined least squares estimates of ϕ were shown to be consistent for any symmetric stable innovations distribution by Kanter and Steiger (1974) and for more general infinite variance distributions by Yohai and Maronna (1977). However, it should also be noted that their results are based on the assumption that the location parameter μ is known.

4. M-ESTIMATES FOR AUTOREGRESSION

The discussion of the previous section suggests that M-estimates for autoregression will be efficiency robust for innovations outlier models. Referring to (2.4) write $z_i^T = (1, y_{i-1}, y_{i-2}, \ldots, y_{i-p})$ and $\beta^T = (\beta_1, \beta_2, \ldots, \beta_{p+1}) = (\gamma, \phi_1, \phi_2, \ldots, \phi_p)$ where $\gamma = \mu[1 - \Sigma_1^p \phi_\ell]$ is the intercept. Then following Huber's (1973) proposal for estimating regression coefficients and residual scale gives

$$\sum_{i=p+1}^{n} z_i \psi\left(\frac{y_i - z_i^T \hat{\beta}_M}{\hat{s}}\right) = 0 \; ; \; \frac{1}{n-2p-1} \sum_{i=p+1}^{n} \psi^2\left(\frac{y_i - z_i^T \hat{\beta}_M}{\hat{s}}\right) = B \qquad (4.1)$$

as the estimating equations for β and the innovations scale s. The constant B is selected so that $\hat{s}$ is consistent for σ_ε when $g = N(0, \sigma_\varepsilon^2)$, i.e., $A = E_\Phi \psi^2(r)$ where Φ is the standard normal distribution. Values of B for a variety of influence functions are given by Martin and Zeh (1978).

Under regularity conditions (Martin, 1978c) consistency and asymptotic normality are obtained for finite variance IO models. The asymptotic variance-covariance matrix for $\hat{\beta}_M$ is

$$V_{\beta,M} = D^{-1} \frac{E\psi^2(\varepsilon)}{E^2\psi'(\varepsilon)} = D^{-1} V_{\ell oc}(\psi,g) \tag{4.2}$$

where $V_{\ell oc}(\psi,g)$ is the usual location M-estimate asymptotic variance (Huber, 1964) at innovations density g and

$$D = Ez_i z_i^T = \left(\begin{array}{c|c} 1 & \mu \underline{1}^T \\ \hline \mu \underline{1} & R \end{array}\right) \tag{4.3}$$

with $\underline{1}$ a px1 vector of 1's and R the pxp moment matrix with elements $R_{ij} = Ey_i y_j$, i,j = 1,2,...,p. Beran (1976) proposed estimating ϕ using a one-step Newton approximation to the solution of the first equation in (4.1), without attempting to estimate the scale parameter s, and obtained the same asymptotic covariance matrix $V_{\beta,M}$.

The usual inversion formula for partitioned matrices yields

$$V_{\beta,M} = \left(\begin{array}{c|c} 1 + \mu^2 \underline{1}^T C^{-1} \underline{1} & -\mu \underline{1}^T C^{-1} \\ \hline -\mu C^{-1} \underline{1} & C^{-1} \end{array}\right) \cdot V_{\ell oc}(\psi,g) . \tag{4.4}$$

The lower-right pxp portion of (4.4) is the covariance matrix $V_{\phi,M}$ of the ϕ estimate, while the 1-1 element is the variance of the intercept. Mann and Wald's (1943) result is the specialization $V_{\beta,LS}$ of (4.4) to the least-squares case where $V_{\ell oc}(\psi,g) = \sigma_\varepsilon^2$, in which case $V_{\phi,LS}$ coincides with (3.2) as it should.

Since $C = \sigma_\varepsilon^2 \tilde{C}$ it is apparent that for heavy-tailed g the M-estimate of ϕ can have far greater precision than the least-squares estimate. For with a good choice of ψ the value of $V_{\ell oc}(\psi,g)$ is relatively stable while σ_ε^2 takes on arbitrarily large values for arbitrarily small heavy-tailed deviations of g from normality. If y_i is a first order autoregression with $\phi = \phi_1$, the asymptotic variance of $\hat{\phi}_M$ is

$$V_{\phi,M} = \frac{1-\phi^2}{\sigma_\varepsilon^2} V_{\ell oc}(\psi,g) \tag{4.5}$$

which provides a particularly transparent special case.

Using (3.3) and (3.4) gives the asymptotic efficiency of $\hat{\phi}_M$ at innovations density g:

$$\text{EFF}(\hat{\phi}_M,g) = \left(\frac{\det I_\phi^{-1}}{\det V_{\phi,M}}\right)^{\frac{1}{p}} = [i(g)\cdot V_{\ell oc}(\psi,g)]^{-1} . \tag{4.6}$$

This is just the pth-power of the asymptotic efficiency of a location M-estimate based on ψ, at an error density g. Thus $\hat{\phi}_M$ has the same attractive asymptotic efficiency robustness as a corresponding location M-estimate using a good psi-function. Some finite sample Monte Carlo results have also been reported (Martin and Denby, 1976; Martin and Zeh, 1978).

Since arbitrarily small positive estimator variances can result from arbitrarily small heavy-tailed deviations of g from normality, the M-estimate $\hat{\phi}_M$ is not asymptotically qualitatively robust as is the least-squares estimate. However, this is hardly a serious deficiency since increased precision is obtained.

It has been assumed that ε_k has a finite variance. I conjecture that $\hat{\beta}_M$ is also consistent for infinite variance ε_k. The variance expressions (4.4) and (4.5) suggest that the ϕ estimates will have a faster rate of convergence than usual. However, a proof is outstanding. The question of the limit distribution in the infinite variance case is likely to be a more difficult one to settle (cf. Logan et al., 1973).

While the IO model is quite tractable and clean asymptotic results are obtained for M-estimates in the finite variance case, the ultimate usefulness of the model is open to question. In the time interval following an isolated innovations outlier ε_{k_o}, the sample path behavior looks very much like the solution to the autoregression homogeneous equation with ε_{k_o} as initial condition. This produces a locally high "signal to noise" ratio which allows for ultraprecise estimation. One may well ask whether or not a model producing such sample paths is as unrealistic as the usual i.i.d. Gaussian assumption. Nonetheless a good reason for retaining the model for awhile is that we can not apriori totally rule out such behavior in the face of the great diversity of time series data types -- and when an innovations outlier does occur in real data it will often be an event of considerable interest (e.g., as in economic time series).

5. ESTIMATING LOCATION WITH AUTOREGRESSIVE ERRORS

The problem now is to obtain robust estimates of the location parameter μ in a pth-order autoregression IO model. For finite-variance autoregressions μ is the mean. Since μ is related to the intercept by $\gamma = \mu[1-\Sigma_1^p \phi_i]$ it is natural to estimate μ with $\hat{\mu} = \hat{\gamma}/[1-\Sigma_1^p \hat{\phi}_i]$ where $\hat{\gamma}$ and the $\hat{\phi}_i$ are robust estimates of γ and ϕ_i, $1 \le i \le p$. If $\hat{\gamma}$ and $\hat{\phi}_i$ are maximum-likelihood estimates then $\hat{\mu}$ is a maximum-likelihood estimate. If $\hat{\beta}^T = (\hat{\gamma}, \hat{\phi}_1, \hat{\phi}_2, \ldots, \hat{\phi}_p)$ is an M-estimate of $\beta^T = (\beta_1, \beta_2, \ldots, \beta_{p+1}) = (\gamma, \phi_1, \phi_2, \ldots, \phi_p)$ then it is appropriate to call $\hat{\mu}$ an (autoregressive errors) M-estimate of μ.

Now $\hat{\alpha}^T = (\hat{\alpha}_1, \hat{\alpha}_2, \ldots, \hat{\alpha}_{p+1}) = (\hat{\mu}, \hat{\phi}_1, \hat{\phi}_2, \ldots, \hat{\phi}_p)$ is an estimate of $\alpha^T = (\alpha_1, \alpha_2, \ldots, \alpha_{p+1}) = (\mu, \phi_1, \phi_2, \ldots, \phi_p)$. The parameters α and β, in conjunction with the innovations distribution, provide equivalent descriptions of the autoregression. Let $\beta = h(\alpha)$ denote the transformation from α to β and let H denote the matrix of partial derivatives of $h(\alpha)$ with respect to α, i.e., $H_{ij} = (\partial/\partial\alpha_j)h_i(\alpha)$. Since $\beta_1 = \alpha_1[1-\Sigma_2^{p+1}\alpha_i]$ and $\beta_i = \alpha_i$, $i=2,\ldots,p+1$ we have

$$H = \left(\begin{array}{c|c} 1-\Sigma_{i=2}^{p+1}\alpha_i & -\alpha_1 \underline{1}^T \\ \hline \underline{0} & I \end{array}\right) = \left(\begin{array}{c|c} 1-\Sigma_1^p \phi_i & -\mu \underline{1}^T \\ \hline \underline{0} & I \end{array}\right) . \tag{5.1}$$

Stationarity insures that $\Sigma_1^p \phi_i < 1$ and hence H is non-singular.

Now the estimate $\hat{\alpha}$ is related to $\hat{\beta}$ by $\hat{\alpha} = h^{-1}(\hat{\beta})$ and expanding about α gives

$$\hat{\alpha} - \alpha = H^{-1} \cdot (\hat{\beta} - \beta) + o(\hat{\beta} - \beta) .$$

If $\hat{\beta}$ is consistent and asymptotically normal with covariance matrix V_β then $\hat{\alpha}$ is consistent and asymptotically normal with covariance matrix

$$V_\alpha = H^{-1} V_\beta (H^T)^{-1} . \tag{5.2}$$

If I_β is the asymptotic information matrix for estimates of β, then $V^*_{CR,\beta} = I_\beta^{-1}$ is the asymptotic lower bound for consistent estimates of β. The information matrix I_α for estimates of α is related to I_β by (Cox and Hinkley, 1974, p. 130; Rothenberg, 1973, p. 31)

$$I_\alpha = H^T I_\beta H \ . \tag{5.3}$$

Thus the Cramer-Rao lower bound for estimates of α is

$$V_{CR,\alpha} = I_\alpha^{-1} = H^{-1} I_\beta^{-1} (H^T)^{-1} \ . \tag{5.4}$$

For the comments to follow assume that the IO autoregression has a finite variance.

If $\hat{\beta}_M$ is an M-estimate then in (5.2) $V_\beta = V_{\beta,M}$ as given by (4.2) and so the asymptotic covariance matrix of the M-estimate $\hat{\alpha}_M$ is

$$V_{\alpha,M} = (H^T DH)^{-1} \cdot V_{\ell oc}(\Psi,g) = \left(\begin{array}{c|c} (1-\Sigma_1^p \phi_i)^{-2} & \underline{0}^T \\ \hline \underline{0} & C^{-1} \end{array}\right) \cdot V_{\ell oc}(\Psi,g) \tag{5.5}$$

with C and $V_{\ell oc}(\Psi,g)$ as in (3.1) and (4.2), respectively.

The least-squares estimate is a special case of M-estimate for which $V_{\ell oc}(\Psi,g) = \sigma_\varepsilon^2$, and the corresponding asymptotic covariance matrix is

$$V_{\alpha,LS} = \left(\begin{array}{c|c} \dfrac{\sigma_\varepsilon^2}{(1-\Sigma_1^p \phi_i)^2} & \underline{0}^T \\ \hline \underline{0} & \tilde{C}^{-1} \end{array}\right) \tag{5.6}$$

with $\tilde{C}$ free of σ_ε^2. This expression places in evidence the distinctively different behavior of the μ and ϕ estimates, the latter being qualitatively robust whereas the former is not. The upper-left element is also the asymptotic variance of the sample mean, as is well known.

The upper-left element of (5.5) differs from usual location M-estimate asymptotic variance $V_{\ell oc}(\Psi,g)$ for i.i.d. errors only by the scale factor $(1-\Sigma_1^p \phi_i)^{-2}$. It follows that Huber (1964) type min-max robustness results hold for (autoregressive-errors) location M-estimates $\hat{\mu}_M = \hat{\gamma}_M/[1-\Sigma_1^p \hat{\phi}_{M,i}]$ over restricted finite-variance families of distributions.

It may be shown (Martin, 1978a) that $I_\beta = D \cdot i(g)$ and so (5.4) gives

$$V_{CR,\alpha} = \frac{1}{i(g)} \begin{pmatrix} (1-\Sigma_1^p \phi_i)^{-2} & \underline{0}^T \\ \underline{0} & C^{-1} \end{pmatrix} . \tag{5.7}$$

The lower-right pxp matrix is the inverse of I_ϕ, the Fisher information for ϕ. The upper-left elements of (5.5) and (5.7) give the efficiency of $\hat{\mu}_M$:

$$\text{EFF}(\hat{\mu}_M, \psi, g) = [i(g) \cdot V_{\ell oc}(\psi, g)]^{-1} \tag{5.8}$$

which is the same as the ordinary location M-estimate efficiency for i.i.d. errors.

If ψ is monotone the autoregressive-errors M-estimate $\hat{\mu}_M$ is almost worthless in additive outlier situations since the same is true of the M-estimates $\hat{\phi}_{M,i}$.[†] The reason for this is given in the next section where I propose an alternative estimate of $\beta^T = (\gamma, \phi^T)$ which is robust toward additive outliers. This alternative estimate can be used to construct a robust estimate of μ. However, the method of construction is somewhat different than that of this section, and will be discussed elsewhere.

6. GM-ESTIMATES FOR AUTOREGRESSION

The last two sections have shown that M-estimates of autoregressive parameters are an attractive possibility for obtaining asymptotic efficiency robustness in situations where only innovations outliers are of concern. When additive outliers occur the above is no longer true. It turns out (Denby and Martin, 1976; Martin and Jong, 1976) that M-estimates have finite sample and asymptotic biases nearly as large as those of least-squares estimates.

As an example of just how bad the latter can be, consider a first order autoregressive AO model

$$x_i = \phi x_{i-1} + \varepsilon_i; \qquad y_i = x_i + v_i$$

where x_i has finite variance σ_x^2, and $\{v_i\}$ is stationary with zero mean, finite variance σ_v^2 and lag-one correlation coefficient ρ. The least-squares estimate $\hat{\phi}_{LS} = (\Sigma_{i=1}^{n-1} y_i y_{i+1}) / (\Sigma_{i=1}^{n-1} y_i^2)$ converges in probability (and almost surely) to ϕ_o where

[†] Although this is not the case for an appropriately-computed M-estimate based on a redescending ψ such as ψ_{BS}, the use of a monotone ψ, such as ψ_H, is needed for such a computation, (cf., Section 1).

$$\phi_o = \phi - (\phi-\rho)\frac{\sigma_v^2}{\sigma_x^2+\sigma_v^2} . \qquad (6.1)$$

Of course $\phi_o = \phi$ as it should be when the v_i are identically zero. But suppose for example that the v_i are uncorrelated and have a common distribution $CND(\delta,\sigma^2) = (1-\delta)N(0,0)+\delta N(0,\sigma^2)$, where $N(\mu,\sigma^2)$ is the normal distribution with mean μ and variance σ^2. Then $P(v_i = 0) = 1-\delta$ and $\sigma_v^2 = \delta\sigma^2$. If $\gamma = .1$, $\sigma^2 = 10$ and $\sigma_x^2 = 1$ then the bias is $-\phi/2$, a 50% asymptotic bias.

The above example shows that $\hat{\phi}_{LS}$ is not qualitatively robust and has a breakdown point of zero. For the distribution $CND(\delta,\sigma^2)$ will be arbitrarily close to the degenerate distribution concentrated at the origin (in any reasonable metric) provided the fraction of contamination $\delta > 0$ is sufficiently small, while $\delta\sigma^2$ can be made arbitrarily large for any given $\delta > 0$ by choosing σ^2 sufficiently large. Correspondingly ϕ_o is arbitrarily close to zero. It is to be noted that patchy additive outliers with $\rho > \phi$ will result in positive bias. Hence $\hat{\phi}_{LS}$ could also be totally ruined with $\phi_o \to 1$ by virtue of $\rho \to 1$ and $\sigma^2 \to \infty$.

Since M-estimates also lack qualitative robustness and have zero breakdown points for AO departures from perfectly observed autoregressive models, a robust alternative is needed. The generalized M-estimates (GM-estimates) defined below are one possibility for obtaining robust estimates of autoregressive models when either innovations or additive outliers are present. The basic idea behind GM-estimates is to modify the M-estimate equations (4.1) so that the summands of the estimating equations are bounded and continuous functions of the data. This in turn results in an influence curve which has the same properties. The comments in Section 1 on bounded-influence regression and the fact that the AO model presents an errors-in-both-variables regression problem are pertinent here.

Let $\hat{\mu}$ be a robust location estimate, computed in a manner to be discussed shortly, and set $\tilde{z}_i^T = (y_{i-1}-\hat{\mu}, y_{i-2}-\hat{\mu}, \ldots, y_{i-p}-\hat{\mu})$. Also let $\hat{C}^{-1}$ be a non-negative definite robust estimate of C^{-1} where C is the pxp covariance matrix for the pth order autoregression x_i. Then set

$$W(\tilde{x}_i) = w(p^{-1}\, \tilde{z}_i^T\, \hat{C}^{-1}\, \tilde{z}_i) \tag{6.2}$$

where $w(\cdot)$ is a non-negative continuous function such that $W(\tilde{z}_i)z_i$ is bounded. Then if ψ is bounded and continuous, the same is true of the summands in the following equations which define a GM-estimate $\hat{\beta}_{GM} = (\hat{\gamma}, \hat{\phi}_1, \ldots, \hat{\phi}_p)$ of β:

$$\sum_{i=p+1}^{n} W(\tilde{z}_i) z_i \psi\left(\frac{y_i - z_i^T \hat{\beta}_{GM}}{\hat{s}}\right) = 0 \;;$$

$$\frac{1}{n-2p-1} \sum_{i=p+1}^{n} W(\tilde{z}_i)\psi^2\left(\frac{y_i - z_i^T \hat{\beta}_{GM}}{\hat{s}}\right) = A \;. \tag{6.3}$$

If $\hat{\mu} = \mu$ and $A = E_\Phi W(\tilde{x}_i)\cdot E_\Phi \psi^2(r)$ then $\hat{s}$ is consistent for σ_ε when a Gaussian IO model holds. Good approximate solutions of these equations (also called GM-estimates) may be obtained using an iterated weighted least-squares (IWLS) technique similar to that mentioned by Huber (1977).

One way of obtaining the $\hat{\mu}$ used in the $\tilde{z}_i$ would be to set $\hat{\mu} = \hat{\mu}_{GM} = \hat{\gamma}/[1-\Sigma_1^p \hat{\phi}_i]$ in (6.3), as is suggested by Section 5, and compute $\hat{\mu}_{GM}$ iteratively during the IWLS procedure. Another possibility is to use an ordinary M-estimate for $\hat{\mu}$. This approach has the advantage that it facilitates arguments concerning the existence and uniqueness of solutions to (6.3). An even simpler possibility is to work totally with data which is robustly centered via ordinary location M-estimates, replacing the z_i by $\tilde{z}_i$ and $\hat{\beta}_{GM}$ by $\hat{\phi}_{GM}$ in (6.3).

The latter approach, which sidesteps direct estimation of the intercept, is the one I have used for computation as of this writing. Some Monte Carlo results for first order situations indicate that this typically results in little loss of efficiency in estimating ϕ relative to estimates based on exact knowledge of μ.

As for the robust estimate $\hat{C}^{-1}$, two possibilities based on alternative representations of C^{-1} deserve mention. The first approach is to express C^{-1} as a function of ϕ, $C^{-1} = C^{-1}(\phi)$, using Siddiqui's (1958) results and then set $\hat{C}^{-1} = C^{-1}(\hat{\phi}_{GM})$. Use of this method appears to create extreme difficulties in establishing existence and uniqueness of solutions to the estimating equations.

The second possibility is based on the factorization $C^{-1} = A^T A$ where A depends only upon the coefficients $\underline{\phi}_i^T = (\phi_{i1}, \phi_{i2}, \ldots, \phi_{ii})$ of the minimum-mean-squared-error linear predictions $\hat{y}_i = \Sigma_{\ell=1}^{i} \phi_{i\ell} y_{k-\ell}$, and the associated minimum-mean-squared errors σ_i^2 for $i = 1,2,\ldots,p-1$ (see Akaike, 1969). Thus C^{-1} is robustly estimated by $\hat{C}^{-1} = \hat{A}^T\hat{A}$ where $\hat{A} = A(\{\hat{\underline{\phi}}_i\}_{i=1}^{p-1}, \{\hat{\sigma}_i\}_{i=1}^{p-1})$, the $\hat{\underline{\phi}}_i$ and $\hat{\underline{\sigma}}_i$ being robust estimates of $\underline{\phi}_i$ and σ_i obtained during the process of fitting autoregressions of orders 1 through p-1. This approach is computationally convenient when one is fitting autoregressive models for a succession of orders, with a view toward selecting a good order by some criterion or another. Furthermore, since $\hat{C}^{-1}$ is fixed when (6.3) is being solved, the existence and uniqueness problem is much easier to solve.

It can be shown that GM-estimates are consistent under innovations outlier models (without Gaussian or finite variance assumptions). Furthermore the GM-estimates are qualitatively robust toward additive outliers and have positive breakdown points under reasonable assumptions. It is somewhat disappointing that the breakdown point is bounded above by $1/(p+1)$ just as in the case of Maronna's robust covariance estimates (1976). Fortunately the effective breakdown point will be higher when only patchy additive outliers occur. Further details are given by Martin (1978d). Small-sample efficiency robustness properties of GM-estimates for both IO and AO situations are reported in Denby and Martin (1976) and Martin and Zeh (1979). See also my paper "Robust estimation of autoregressive models" (to appear in the Proceedings of the Instit. for Math. Stat. Special Conference on Time Series, Ames, Iowa, May 1-3, 1978). The latter paper contains further details about GM-estimates, along with some application examples and a proposal for a robust order-selection rule.

7. CONDITIONAL MEAN M-ESTIMATES

The estimates introduced in this section appear to be natural versions of M-estimates for time series with additive outliers. Consider an autoregressive AO model (2.1) and assume that $\mu = 0$ (translation equivariant versions of the estimates are obtained in a straightforward manner). Let $Y_i^T = (y_1, \ldots, y_i)$, $\underline{x}_i^T = (x_i, x_{i-1}, \ldots, x_{i-p+1})$, and for short let $\hat{\underline{x}}_i(\phi')$ denote the conditional mean estimate $E\{\underline{x}_i | Y_i, \phi'\}$ of $\underline{x}_i$ under the assumption that ϕ' is the autoregressive parameter vector.

A conditional-mean M-estimate (CMM-estimate) is a solution of the minimization problem

$$\min_{\phi'} \sum_{i=p}^{n-1} \rho\left(\frac{y_{i+1} - \hat{\underline{x}}_i^T(\phi')\phi'}{\hat{s}}\right) \tag{7.1}$$

where ρ is a symmetric robustifying loss function and the scale estimate $\hat{s}$ is yet to be specified. Since the solution is a stationary point we have

$$\sum_{i=p}^{n-1} [\hat{\underline{x}}_i(\hat{\phi}) + D_i(\hat{\phi})\hat{\phi}] \cdot \psi\left(\frac{y_{i+1} - \hat{\underline{x}}_i^T(\hat{\phi})\hat{\phi}}{\hat{s}}\right) = 0 \tag{7.2}$$

where

$$[D_i(\hat{\phi})]_{kj} = \left.\frac{\partial\, \hat{x}_{i-j+1}(\phi')}{\partial\, \phi'_k}\right|_{\phi' = \hat{\phi}}, \quad k,j = 1,\ldots,p\,. \tag{7.3}$$

The estimating equation (7.2) is rather unwieldy due to the presence of $D_i(\hat{\phi})$. There is some evidence, in the form of both heuristic arguments and Monte Carlo, that the terms $D_i(\hat{\phi})\cdot\hat{\phi}$ may be dropped without seriously degrading the estimate, and so we turn to the simpler approximate version

$$\sum_{i=p}^{n-1} \hat{\underline{x}}_i(\hat{\phi})\cdot\psi\left(\frac{y_{i+1} - \hat{\underline{x}}_i^T(\hat{\phi})\hat{\phi}}{\hat{s}}\right) = 0 \quad . \tag{7.4}$$

The resulting $\hat{\phi}$ has the appealing feature that it is obtained by robustly regressing the observations on conditional-mean estimates $\{\hat{\underline{x}}_i(\hat{\phi})\}$ of the (occasionally) unobservable $\{x_i\}$.

On way to obtain the estimate $\hat{s}$ would be by adjoining the additional equation

$$\frac{1}{n-2p} \sum_{i=p}^{n-1} \psi^2\left(\frac{y_{i+1}-\hat{\underline{x}}_i^T(\hat{\phi})\hat{\phi}}{\hat{s}}\right) = B \tag{7.4'}$$

as is suggested by (4.1). Another method will be mentioned shortly.

In order to solve (7.4) we need to express $\hat{\underline{x}}_i(\phi') = E\{\underline{x}_i|Y_i,\phi'\}$ as a function of ϕ' and Y_i for the pth-order autoregressive AO model. Masreliez's (1975) results indicate that a good approximate version of the estimates $\hat{\underline{x}}_i = E\{\underline{x}_i|y_i,\phi'\}$ may be computed recursively by a <u>robust filter with time-varying data-dependent scaling</u>. With

$$m_i = \begin{pmatrix} m_{1i} \\ m_{2i} \\ \cdot \\ \cdot \\ \cdot \\ m_{pi} \end{pmatrix} \quad \text{and} \quad \Phi' = \begin{pmatrix} \phi_1' & \phi_2' & \cdot & \cdot & \cdot & \cdot & \phi_p' \\ 1 & 0 & \cdot & \cdot & \cdot & \cdot & 0 \\ 0 & 1 & 0 & \cdot & \cdot & \cdot & 0 \\ \cdot & & & & & & \cdot \\ \cdot & & & & & & \cdot \\ \cdot & & & & & & \cdot \\ 0 & 0 & \cdot & \cdot & 0 & 1 & 0 \end{pmatrix}$$

the filter recursion is

$$\hat{\underline{x}}_{i+1} = \Phi'\hat{\underline{x}}_i + m_i \cdot \Psi_i(y_{i+1}-\hat{\underline{x}}_i^T\phi') \tag{7.5}$$

where $\hat{\underline{x}}_i^T\phi'$ is the one-step-ahead prediction of both x_{i+1} and y_{i+1}, and m_i is computed by an auxiliary data-dependent recursion formula. The function $\Psi_i(y_{i+1}-\hat{\underline{x}}_i^T\phi')$ of y_{i+1} (with $\underline{x}_i^T\phi'$ regarded as fixed) is the score function for the observation prediction density $f(y_{i+1}|Y_i,\phi')$, i.e., $\Psi_i(y_{i+1}-\hat{\underline{x}}_i^T\phi') = -(\partial/\partial y_{i+1})\ln f(y_{i+1}|Y_i,\phi')$, computed under the assumption that the x-prediction density $f(x_{i+1}|Y_i,\phi')$ is Gaussian. The subscript on Ψ_i reflects the fact that Ψ_i contains a time-varying data-dependent scale factor which depends upon m_{1i}.

When the v_i are identically zero the x-prediction density is in fact Gaussian and $m_i \cdot \Psi_i(y_{i+1}-\hat{\underline{x}}_i^T\phi') = \eta \cdot (y_{i+1}-\hat{\underline{x}}_i^T\phi')$ where $\eta^T = (1,0,0,\ldots,0)$. Thus $\hat{\underline{x}}_i^T = (y_i,y_{i-1},\ldots,y_{i-p+1})$ for $i \geq p$, and the filter is the identity operator as one would expect.

Since we will not know the distribution of the v_i and wish to have protection against distributions of the form $F_v = (1-\delta)N(0,0) + \delta H$, for a variety of H, we need to replace $m_i \cdot \Psi_i(y_{i+1}-\hat{\underline{x}}_i^T\phi')$ with $m_i \cdot \psi_i(y_{i+1}-\hat{\underline{x}}_i^T\phi')$ where ψ_i is a bounded and continuous function. Computation of Ψ_i for the case

$F_v = \mathrm{CND}(\delta,\sigma^2) = (1-\delta)N(0,0) + \delta N(0,\sigma^2)$ suggests to use the form $\psi_i(t) = \hat{s}_i^{-1/2} \cdot \psi(t \cdot \hat{s}_i^{-1/2})$ where $\hat{s}_i^2 = m_{1i}$ is the conditional prediction error variance for predicting x_{i+1} given Y_i. Further details, including some Monte Carlo results for various ψ shapes are given by Martin and DeBow (1976).

A possible design criterion for a robust filter of the form (7.5), with Ψ_i replaced by ψ_i, is the following. The filter should be nearly the identity operator in some sense. For example use of Mallow's (1977) interesting results on the linear part of non-linear smoothers would lead to the criterion: the linear part should be nearly the identity operator. With such criteria in mind, one would choose ψ shapes for which $\psi'(t) = 1$ on $[-K,K]$ where K is chosen as large as possible, consistent with obtaining sufficient robustness by bounding the influence of large prediction residuals.

An attractive simplification of the robust filter (7.5) with $\Psi_i = \psi_i$ and $m_{1i} = \hat{s}_i^2$ is

$$\hat{x}_{i+1} = \hat{\underline{x}}_i^T \phi' + \hat{s}_i \, \psi\left(\frac{y_{i+1} - \underline{x}_i^T \phi'}{\hat{s}_i}\right) \tag{7.6}$$

with $\hat{s}_i$ obtained from the appropriate data-dependent auxiliary recursion. Presumably this simplification entails some loss in efficiency.

A further simplication of (7.6) is the robust filter

$$\hat{x}_{i+1} = \hat{\underline{x}}_i^T \phi' + \hat{s} \, \psi\left(\frac{y_{i+1} - \underline{x}_i^T \phi'}{\hat{s}}\right) \tag{7.7}$$

where $\hat{s}$ is a data-dependent but time-invariant (i.e., globally-determined) estimate of scale for the prediction residuals $y_{i+1} - \underline{x}_i^T \phi'$. For example $\hat{s}$ might be determined by (7.4) - (7.4'). Some theoretical and empirical justifications for filters of the above form have been given by Masreliez and Martin (1977). Robust filters of this variety were also introduced independently and quite successfully by D. J. Thomson (1977) in connection with spectral density estimation.

It is not yet entirely clear whether or not the simplification (7.7) is a good idea. I have a strong preference for (7.6) because its structure is more closely related to that of the approximate conditional mean. As a result the scale factor $\hat{s}_i$ depends upon the local character of the data in an intuitively appealing manner (see Martin and DeBow, 1976). Furthermore, I would add a word of warning. If ψ is of the non-monotone redescending variety then the version (7.7) is unsafe (for it can then loose track of the data, never to regain it), and use of either (7.5) or (7.6) is generally imperative.

Notice that if ψ and $\hat{s}$ are chosen to be the same in (7.4) and (7.7), which is hardly unreasonable, then it is not necessary to solve equation (7.4) directly. For multiplying both sides of (7.7) by $\hat{\underline{x}}_i = \hat{\underline{x}}_i(\phi')$ and summing on i shows that (7.4) is equivalent to the Yule-Walker type normal equation

$$\sum_{i=p}^{n-1} \hat{\underline{x}}_i(\hat{\phi})\cdot[\hat{x}_{i+1}(\hat{\phi}) - \hat{\underline{x}}_i^T(\hat{\phi})\hat{\phi}] = 0 \quad . \tag{7.8}$$

The above equation invites the iterative solution

$$\sum_{i=p}^{n-1} \hat{\underline{x}}_i(\hat{\phi}^j)\cdot[\hat{x}_{i+1}(\hat{\phi}^j) - \hat{\underline{x}}_i^T(\hat{\phi}^j)\hat{\phi}^{j+1}] = 0 \qquad j = 1,2,\ldots,\text{NIT} \tag{7.9}$$

where $\hat{\underline{x}}_i(\hat{\phi}^j)$ is obtained from (7.7) with $\phi' = \hat{\phi}^j$, and $\hat{\phi}^1$ is the least-squares estimate.

Notice that if the variable scale filter (7.6) is used instead of the fixed scale filter (7.7) then replacement of $\hat{s}$ by $\hat{s}_i$ in (7.4) again leads to the Yule-Walker type equation (7.8). If $\hat{s}$ is replaced by $\hat{s}_i$ in (7.1), then an approximate model AO maximum-likelihood estimate is obtained. Although the detailed nature of this approximation is not understood, it nonetheless provides some additional motivation for the use of CMM-estimates.

When the observed series $\{y_i\}$ contains a relatively small fraction of outliers, the properly calibrated robust scale estimate $\hat{s}$ computed from (7.4') should differ relatively little from the square root of the usual s^2 computed from the residuals of the final iteration of (7.9). Thus the latter simpler method, which fits in well with conventional least-squares procedures, may be adequate.

A number of important theoretical questions related to CMM-estimates remain to be explored and considerable effort will probably be required to understand the nature of the various approximations mentioned here. However, the basic idea behind the CMM-estimate is quite attractive, and the version (7.8) of the estimating equation has considerable intuitive appeal. For suppose ψ is of the redescending rejection variety. Then the estimation procedure corresponds to replacing outliers which are sufficiently large on the scale of the prediction residuals $\{y_{i+1}-\hat{\underline{x}}_i^T(\hat{\phi})\hat{\phi}\}$ by predicted values (and to intermediate treatment of not-so-obvious outliers), and then using least-squares on the modified data.

Some exploratory Monte Carlo results for the case of first-order autoregression with $\phi = .5$ and $\phi = .9$ yielded smaller biases (appreciably smaller at $\phi = .9$) for CMM-estimates than GM-estimates at non-Gaussian AO situations. The corresponding variances were also typically smaller. Efficiencies at the Gaussian situation were reasonably high. Although these results were in the direction expected, much more thorough studies are required before firm conclusions may be drawn. It should also be mentioned that the exploratory Monte Carlo also showed (not unexpectedly) that CMM-estimates can do quite poorly at heavy-tailed IO situations.

A final remark is that the conditional mean $\hat{\underline{x}}_i = E\{\underline{x}_i | Y_i, \phi'\}$ in (7.1) might well be replaced with the conditional mean $\hat{\underline{x}}_i = E\{\underline{x}_i | Y_n, \phi'\}$. The latter depends upon all of the observed data, and is termed a "smoother". This suggests to use a robust smoother in place of a robust filter to obtain the $\hat{\underline{x}}_i$ in equation (7.8). Preliminary Monte Carlo studies show, not unexpectedly, that the resulting estimates of ϕ are better than those based on a robust filter. Detailed results will be presented elsewhere.

8. DETERMINING OUTLIER TYPE

Although GM-estimates appear to do moderately well on an overall basis at both innovations and additive outlier situations, M-estimates are clearly superior when only innovations outliers occur (Denby and Martin, 1976; Martin and Zeh, 1978). Approximate versions of CMM-estimates promise to yield better results than GM-estimates at AO situations, but as was mentioned previously they can perform poorly at non-Gaussian IO situations. Thus methods for ascertaining which type of outliers are present in a given data set would be desirable as guidance in selecting an estimator, and for other reasons as well. I shall briefly describe two possibilities. One is a significance test and the other is a residuals-plotting technique.

Suppose we wish to test the composite null hypothesis H_{IO} that a Gaussian or non-Gaussian IO model holds versus the composite alternative H_{AO} that additive outliers are present. Then since M-estimates have large biases while GM-estimates have relatively small biases for AO situations, a significance test based on the difference $\hat{\phi}_M - \hat{\phi}_{GM}$ between an M-estimate and a GM-estimate suggests itself.

It can be shown that under reasonable assumptions the asymptotic distribution of $\delta_n = \sqrt{n}(\hat{\phi}_{M,n} - \hat{\phi}_{GM,n})$ is multivariate normal and under H_{IO} has mean zero and covariance matrix

$$V_{\delta,IO} = [B_1^{-1} B_2 B_1^{-1} - C^{-1}] \cdot V_{\ell oc} \qquad (8.1)$$

where

$$B_1 = E\tilde{y}_p W(\tilde{y}_p)\tilde{y}_p^T, \quad B_2 = E\tilde{y}_p W^2(\tilde{y}_p)\tilde{y}_p^T, \quad C = E\tilde{y}_p\tilde{y}_p^T,$$

$$\tilde{y}_p^T = (y_p - \mu, y_{p-1} - \mu, \ldots, y_1 - \mu)$$

and all expectations are taken under H_{IO}. Thus the asymptotic distribution of $T_n = \delta_n^T V_{\delta,IO}^{-1} \delta_n$ is chi-squared with p degrees of freedom. A usable test statistic might be obtained by replacing $V_{\delta,IO}$ by a good estimate $\hat{V}_{\delta,IO}$ and using χ_p^2 critical values, or perhaps critical values obtained via Monte Carlo. Further details and some encouraging Monte Carlo results for first order autoregression are given by Martin and Zeh (1977).

The scatter plot approach for assessing outlier type in an exploratory manner is based on the residuals $r_i = y_{i+1} - \Sigma_{\ell=1}^p \hat{\phi}_\ell \, y_{i-\ell+1}$, $p \le i \le n-1$, from a GM-estimate fit. If the IO model holds then $r_i \approx \varepsilon_{i+1}$, $p \le i \le n-1$, for a good pth-order fit. Since the ε_i are independent a scatter plot of the pairs (r_i, r_{i+1}) will reveal outliers lying mainly along the abscissa and ordinate when ε_i is heavy-tailed. If the AO model holds then $r_i \approx \varepsilon_{i+1} + v_{i+1} - \Sigma_{\ell=1}^{p} \phi_\ell \, v_{i-\ell+1}$ and so the residuals have a correlation structure attributable to the outliers. In this case the outliers generally no longer lie mainly along the abscissa and ordinate in the scatter plot. The distinctively different characters of the outlier configurations in scatter plots under H_{IO} and H_{AO} may be used as exploratory indicators of outlier type. Some examples for both artificial and real data sets are given in Martin and Zeh (1977).

9. ACKNOWLEDGEMENT

I have benefited considerably from many discussions with Lorraine Denby, Beat Kleiner and Dave Thomson at Bell Laboratories, and James Jong and Judy Zeh at the University of Washington. We have been co-workers on various aspects of robust estimation for time series, some of which are reported in this paper and reflected in the references.

REFERENCES

1. Abraham, B. and Box, G. E. P. (1975), "Outliers in time series," Tech. Report No. 440, Dept. of Statist., University of Wisconsin.

2. Akaike, H. (1969), "Power spectrum estimation through autoregressive model fitting," Annals Instit. Statist. Math., 21, 407-419.

3. Anderson, T. W. (1971), The Statistical Analysis of Time Series, John Wiley.

4. Andrews et al. (1972), Robust Estimates of Location-Survey and Advances, Princeton University Press.

5. Beran, R. (1976), "Adaptive estimates for autoregressive processes," Annals Instit. Statist. Math., 28, 77-89.

6. Cox, D. R. and Hinkley, D. V. (1974), Theoretical Statistics, Halstead Press (John Wiley), New York.

7. Denby, L. and Martin, R. D. (1976), "Robust estimation of the first order autoregressive parameter," Bell Labs Tech. Memo, Murray Hill, N.J. (to appear in Jour. Amer. Stat. Assoc.).

8. Denby, L. and Larsen, W. (1977), "Robust regression estimators compared via Monte Carlo," Comm. in Statist., Vol. A6, No. 4, 335-362.

9. Devlin, S. J., Gnanadesikan, R. and Kettenring, J. R. (1975), "Robust estimation and outlier detection with correlation coefficients," Biometrika, 62, No. 3.

10. Dzhaparidze, K. O. and Yaglom, A. M. (1973), "Asymptotically efficient estimation of the spectrum parameters of stationary stochastic processes," Proceedings of the Prague Symp. on Asymp. Statist., Charles University Press, Prague.

11. Fox, A. J. (1972), "Outliers in time series," Jour. Roy. Statist. Soc., B, 34, 350-363.

12. Gastwirth, J. L. and Rubin, H. (1975), "The behavior of robust estimators on dependent data," Annals Statist., 3, No. 5, 1070-1100.

13. Hampel, F. (1968), "Contributions to the theory of robust estimation," Ph.D. dissertation, Dept. of Statist., University of California, Berkeley.

14. Hampel, F. (1971), "A general qualitative definition of robustness," Annals Math. Stat., 42, 1887-1896.

15. Hampel, F. (1973), "Robust estimation: a condensed partial survey," Z. Wahrscheinlichkeitstheorie verw. Geb. 27, 87-104.

16. Hampel, F. (1974), "The influence curve and its role in robust estimation," Jour. Amer. Statist. Assoc., 69, No. 346, 383-393.

17. Hampel, F. (1975), "Beyond location parameters: robust concepts and methods," Proceedings of I.S.I. meeting, 40th session, Warsaw.

18. Hill, R. (1977), "Robust regression when there are outliters in the carriers," Ph.D. dissertation, Dept. of Statist., Harvard University.

19. Hogg, R. V. (1974), "Adaptive robust procedures: a partial review and some suggestions for future applications and theory," J. Amer. Statist. Assoc., 69, 909-923.

20. Hogg, R. V. (1978), "An introduction to robust estimation," to appear in Robustness in Statistics, ed. by R. L. Launer and G. Wilkinson, Academic Press.

21. Huber, P. J. (1964), "Robust estimation of a location parameter," Annals Math. Stat., 35, 73-101.

22. Huber, P. J. (1973), "Robust regression: asymptotics, conjectures and Monte Carlo," Annals of Statist., 1, No. 5, 799-821.

23. Huber, P. J. (1977), Robust Statistical Procedures, Society for Industrial and Applied Mathematics, Philadelphia.

24. Kanter, M. (1972), "Linear sample spaces and stable processes," Jour. Func. Anal., 9, 441-459.

25. Kanter, M. and Steiger, W. L. (1974), "Regression and autoregression with infinite variance," Advances in Applied Prob., 6, 768-783.

26. Kazakos, P. and Gray, R. M. (1977), "Robustness of estimators on stationary observations," Bell Labs. Tech. Memo., Holmdel, N.J. (to appear in Annals. of Prob.).

27. Kleiner, R., Martin, R. D. and Thomson, D. J. (1977), "Robust estimation of power spectra," Bell Labs. Tech. Memo., Murray Hill, N.J. (to appear in Jour. Roy. Statist. Soc. B, with discussion).

28. Logan, B. F., Mallows, C. L., Rice, S. O. and Shepp, L. A. (1973), "Limit distributions of self-normalized sums," Annals. of Prob., 1, No. 5, 788-809.

29. Mallows, C. (1976), "On some topics in robustness," Bell Labs. Tech. Memo., Murray Hill, N.J. (talks given at NBER Workshop on Robust Regression, Cambridge, Mass., May, 1973, and at the ASA-IMS Regional Meeting, Rochester, N.Y., May 21-23, 1975).

30. Mallows, C. (1977), "Some theory of non-linear smoothers," Bell Labs. Tech. Memo., Murray Hill, N.J. (talk given at IMS Eastern Regional Meeting, Chapel Hill, N.C., April 8, 1977).

31. Mann, H. B. and Wald, A. (1943), "On the statistical treatment of linear stochastic difference equations," Econometrica, 11, Nos. 3 & 4, 173-220.

32. Maronna, R. (1976), "Robust M-estimators of multivariate location and scatter," Annals of Statist., 4, No. 1, 54-67.

33. Maronna, R. A. and Yohai, V. J. (1977), "Robust M-estimators for regression with contaminated independent variables," unpublished manuscript.

34. Martin, R. D. (1978a), "The Cramer-Rao bound and robust M-estimates for autoregression," Tech. Rep. No. 210, Dept. of Elec. Engrg., University of Washington, Seattle.

35. Martin, R. D. (1978b), "Robust estimates of the mean with autoregressive errors," Tech. Rep. No. 211, Dept. of Elec. Engrg., University of Washington, Seattle.

36. Martin, R. D. (1978c), "Asymptotic properties of M-estimates for pth-order autoregressions," Tech. Rep. No. 212, Dept. of Elec. Engrg., University of Washington, Seattle.

37. Martin, R. D. (1978d), "Asymptotic properties of generalized M-estimates for autoregressive parameters," Tech. Rep. No. 213, Dept. of Elec. Engrg., University of Washington, Seattle.

38. Martin, R. D. and DeBow, G. (1976), "Robust filtering with data-dependent covariance," Proceedings of the Johns Hopkins Conference on Information Sciences and Systems, March 31-April 2.

39. Martin, R. D. and Jong, J. (1976), "Asymptotic properties of robust generalized M-estimates for the first order autoregressive parameter," Bell Labs. Tech. Memo., Murray Hill, N.J.

40. Martin, R. D. and Zeh, J. E. (1977), "Determining the character of time series outliers," Proceedings of the Amer. Statist. Assoc., Business and Economics Section.

41. Martin, R. D. and Zeh, J. E. (1978), "Robust generalized M-estimates for autoregressive parameters: small-sample behavior and applications," Tech. Rep. No. 214, Dept. of Elec. Engrg., University of Washington, Seattle.

42. Masreliez, C. J. (1975), "Approximate non-Gaussian filtering with linear state and observation relations," IEEE Trans. on Auto. Control., AC-20, 107-110.

43. Masreliez, C. J. and Martin, R. D. (1977), "Robust Bayesian estimation for the linear model and robustifying the Kalman filters," IEEE Trans. on Auto. Control, AC-22, 361-371.

44. Mosteller, F. and Tukey, J. W. (1977), Data Analysis and Regression, Addison-Wesley.

45. Relles, D. A. (1968), "Robust regression by modified least squares," Ph.D. dissertation, Dept. of Statist., Yale University.

46. Rothenberg, T. J. (1973), Efficient Estimation with A Priori Information, Cowles Foundation for Research in Economics and Statistics at Yale University, Monograph 23, Yale University Press.

47. Siddiqui, M. M. (1958), "On the inversion of the sample covariance matrix in a stationary autoregressive process," Ann. Math. Statist., 29, 585-588.

48. Thomson, D. J. (1977), "Spectrum estimation techniques for characterization and development of WT4 waveguide-I," Bell System Tech. Jour., 56, No. 4, 1769-1815.

49. Tukey, J. W. (1960), "A survey of sampling from contaminated distributions," in Contributions to Probability and Statistics (ed. Olkin), Stanford University Press.

50. Whittle P. (1953), "Estimation and information in stationary time series," Arch. Math. 2, 423-434.

51. Whittle, P. (1962), "Gaussian estimation in stationary time series," Bull. Int. Stat. Inst., 39, 105-129.

52. Wold, H. (1953), A Study in the Analysis of Stationary Time Series, Stockholm.

53. Yohai, V. J. and Maronna, R. A. (1977), "Asymptotic behavior of least-squares estimates for autoregressive processes with infinite variances," Annals of Statist., 5, No. 3, 554-560.

This work was supported in part by NSF Grant ENG76-00504.

Departments of Electrical Engineering
and Biostatistics
University of Washington
Seattle, Washington 98195

Consultant, Bell Laboratories
Murray Hill, N.J. 07947

Robust Techniques in Communication

V. David VandeLinde

One of the tasks of an engineer is to adapt ideas and results from other disciplines to his problems. The research of the last decade on statistical robustness provides a course of inquiry we have been attempting to focus on the problems faced by the communications engineer. In this paper we concentrate on two areas where robust ideas have been applied: estimation, and detection. Throughout the paper, we use the minmax notion of robustness. We will be searching for the distribution least favorable for the class of distributions specified by our partial knowledge of the state of nature and correspondingly the statistical procedure providing the greatest lower bound on performance over the admissible distributions.

I. ESTIMATION

The observations are of the form

$$x_n = \theta + v_n \ , \ n \geq 1 \ ,$$

where the v_n are independently distributed random variables, each with pdf G which is symmetric about zero. Available information on G is incomplete and so it is used to define a convex set C of symmetric pdf's, each with zero location parameter, to which G is confined. An estimate T is defined as a sequence $\{T_n\}$ of functions $T_n : R^n \to R$, where R is the real line. If F is in C and T is an estimate for which $T_N(\overline{x}_n) \to \theta$ almost surely or in probability and $n^{.5}T_n(\overline{x}_n)$ is

ISBN 0-12-438150-2

asymptotically normal when the v_n are distributed as $F(\bar{x}_n=(x_1,\ldots, x_n))$, denote the asymptotic variance as $V[T,F]$. We seek a solution for the following problem.

Problem 1: Find a pair $[T_0,F_0]$ such that

$$V[T_0,F] \leq V[T_0,F_0] \leq V[T,F_0]$$

for F in C and T a regular, translation invariant estimate.

An estimate T defined by the sequence $\{T_n\}$ is regular if T_n is measurable for each n. The estimate T_0 minimizes the asymptotic variance maximized over C. If T_0 is used to estimate θ, the asymptotic variance of $T_n(\bar{x}_n)$ will be no greater than $V[T_0,F_0]$ regardless of which distribution from C happens to be G.

Let I be the functional which maps each pdf onto its Fisher information for location ($I(F) = \int (f'/f)^2 f\,dx$ if F has an absolutely continuous density f and $I(F) = \infty$ otherwise). Throughout the paper we shall use a prime to denote the derivative with respect to the argument and a lower case letter to indicate the density for the pdf having the corresponding upper case symbol. Let F_0 be a pdf which minimizes I over C, and put $\psi_0 = -f_0'/f_0$. Huber has shown that if the sequence of functions defining T_0 is given by

$$\sum_{i=1}^{n} \psi_0\,(x_i - T_n(\bar{x}_n)) = 0\ , \tag{1}$$

then, provided some regularity conditions on F_0 and C are met, the pair $[T_0,F_0]$ solves Problem 1. The estimate defined by (1) is the maximum likelihood estimate for the least favorable pdf F_0. Huber [2] called an estimate of this form an M-estimate. Note that the M-estimate solving Problem 1 is completely specified once the distribution F_0 of minimum Fisher information in C is known.

I.1 THE ROBBINS-MONRO STOCHASTIC ALGORITHM AS A ROBUST LOCATION ESTIMATE

In this section we show that in general, subject to regularity conditions, an estimate based on the Robbins-Monro stochastic approximation algorithm exists which solves Problem 1. This result corresponds to that given by Huber for M-estimates. A stochastic approximation (SA) estimate solving a special case of Problem 1 was first discovered by

Martin [8]. He gave an SA solution for the case in which the distribution set C is in the ε-contaminated normal class. Later, he and Masreliez [9] found an SA solution for C in what they called the p-point class. Our results show how to obtain an SA solution for Problem 1 for a general distribution set C.

The SA-estimates have a significant computational advantage over M-estimates. The latter require solution of a non-linear equation (1) which gets increasingly burdensome as the number of observations increases. The SA-estimates are given by a simple recursive formula (the stochastic approximation algorithm) and do not require the explicit use of past observations in computing the current estimate. As pointed out by Martin and Masreliez [9], they are thus suitable for real-time applications. However, the M-estimates will, in general, converge more quickly.

Before establishing the existence of an SA estimate solving Problem 1 in the general case, it will be helpful to review the general solution procedure and to define some conventions and terminology which will permit a more concise statement of the results to follow.

There are three steps involved in finding a solution for Problem 1. These are essentially the same for both the M and SA-estimate solutions. Only the formulas defining the two estimates are different. The three solution steps are:

Step 1: Find a distribution F_0 which has minimum information in set C.

Step 2: Check to see if the estimate T_0 defined from F_0 satisfies some available set of sufficient conditions for asymptotic normality. (General conditions sufficient for asymptotic normality have been given by Huber [3] for M-estimates and by Sacks [12] for SA-estimates.)

A positive result in Step 2 implies that $[T_0, F_0]$ solves Problem 1 over the set

$$C^* = \{F \in C: \ I(F) < \infty \} \qquad (2)$$

Step 3: If the result of Step 2 is affirmative, check to see if the solution can be extended from C^* to all of C.

This solution procedure has been established by Huber [2] for M-estimates. We shall show that it holds as well for SA-estimates. That a positive result in Step 2 implies $[T_0, F_0]$ solves Problem 1 over C^* for T_0 an SA-estimate is shown in theorem 1. Theorems 2 and 3 give conditions which are sufficient for extending the solution from C^* to all of C. These conditions apply to M-estimates as well.

Hereafter, we shall use $T^M(T^{SA})$ to represent an M(SA)-estimate. If F_0 minimizes Fisher information on the set C, we define T_0^{SA} to be the sequence generated by the recursive formula

$$T_{n+1}(\bar{x}_{n+1}) = T_n(\bar{x}_n) + \psi_0(x_{n+1} - T_n(\bar{x}_n))/[(n+1)I(F_0)],$$

$$T_1(x_1) = x_1, \quad (3)$$

where $\bar{x}_n$ and ψ_0 are as previously defined. T_0^M is defined by (1). For each F in C, put

$$v_F = E_F[\psi_0^2(x-\theta)],$$

$$v_0 = E_{F_0}[\psi_0^2(x-\theta)], \text{ and } m_F(\alpha) = E_F[-\psi_0(x-\alpha)].$$

If the sufficient conditions for asymptotic normality given by Huber [3] are satisfied by ψ_0 and F, then

$$V[T_0^M, F] = \frac{v_F}{[m_F'(\theta)]^2} \quad (4)$$

If ψ_0 and F meet the corresponding conditions given by Sacks [12], then

$$V[T_0^{SA}, F] = \frac{v_F}{v_0[2m_F'(\theta) - v_0]} . \quad (5)$$

In the following we use T_0 in statements which apply to both T_0^M and T_0^{SA}. Since only translation invariant estimates are considered, there is no loss in generality from the assumption $\theta=0$ which is made for convenience through the rest of the paper. If the sequence $\{n^{.5}T_n\}$ corresponding to T_0 is asymptotically normal when the observation error variables v_n are distributed as F, we say that T_0 is asymptotically normal at F.

The first theorem gives conditions under which an SA-estimate solves Problem 1 over a set which contains only pdf's which have finite Fisher information.

THEOREM 1: Suppose that C is a convex set of pdf's with C^* non-empty and that F_0 minimizes I over C. If at each F in C^*,

(i) T_0^{SA} is consistent and asymptotically normal with variance given by (5), and

(ii) $m_F'(\theta) = -\int\psi_0(x)f'(x+\theta)\,dx$,

then $[T_0^{SA},F_0]$ solves Problem 1 over C^*.

PROOF: A direct substitution in (5) yields $V[T_0^{SA},F_0] = 1/I(F_0)$, the Cramer-Rao lower bound for F_0. It follows that

$$V[T_0^{SA},F_0] \leq V[T,F_0] \tag{6}$$

for all regular, translation invariant estimates T. From conditions (i) and (ii) we obtain

$$V[T_0^{SA},F] = v_F/v_0[\int(-2\psi_0(f'-f_0') - \psi_0^2(f-f_0))dx+v_F], \tag{7}$$

F in C^*. But since F_0 has minimum Fisher information over C^*, theorem 2 of [2] proves that

$$\int[-2\psi_0(f'-f_0') - \psi_0^2(f-f_0)]\,dx \geq 0,\ F \text{ in } C^*.$$

This inequality, through (7), implies that

$$V[T_0^{SA},F] \leq 1/v_0 = V[T_0^{SA},F_0],\ F \text{ in } C^*. \tag{8}$$

The relations (6) and (8) establish the proof.

Condition (ii) allows Huber's differentiability results for I to be applied which makes the proof a simple matter. Some conditions on ψ_0 which imply condition (ii) are

(a) ψ_0 is absolutely continuous and ψ_0' is uniformly bounded a.e. on R with respect to Lebesgue measure, or the weaker condition,

(b) For each F in C^* there is an F-integrable function h_F and a positive number ζ_F such that

$$|(\psi_0(x)/\alpha)\ [f(x+\alpha) - f(x)]| \leq h_F(x),\ F \text{ a.e., when } |\alpha|<\zeta_F.$$

The stronger and more easily verified condition (a) is not restrictive. The distribution of least Fisher information in C is also the smoothest in a sense. Thus, if C is a relatively full set, ψ_0 will be well behaved and condition (a) will be easily satisfied.

I.2 EXTENDING THE MINMAX ESTIMATE TO LARGER DISTRIBUTION SETS

Now we turn our attention toward extending to all of C a solution for Problem 1 which has already been shown to hold over those distributions in C which have finite Fisher information. Conditions allowing this extension are given which apply to both M and SA-estimates. Since the second inequality in the saddle condition in Problem 1 is verified in obtaining a solution which holds over C^*, to extend this solution to the whole set C only requires verification that the first inequality of the saddle condition holds over C. The two theorems to follow give conditions which are sufficient to establish this first inequality.

The next theorem gives a condition for checking to see if the distribution set C is contained in the largest set (the set of all pdf's which satisfy (9)) over which

$$V[T_0^{SA},F] \leq V[T_0^{SA},F_0].$$

If the answer is affirmative then also

$$V[T_0^{M},F] \leq V[T_0^{M},F_0] \text{ over C.}$$

<u>THEOREM 2</u>: Assume that F_0 has minimum Fisher information in C, T_0^M (T_0^{SA}) is asymptotically normal with mean θ and positive finite variance given by (4) ((5)), and $m_F'(\theta) = \int\psi_0'(x-\theta)dF(x)$ for each F in C. Then $[T_0^M,F_0]$ solves Problem 1 if ($[T_0^{SA},F_0]$ solves Problem 1 if and only if

$$\int[2\psi_0'(x-\theta) - \psi_0^2(x-\theta)]d(F(x) - F_0(x)) \geq 0 \tag{9}$$

for each F in C.

<u>PROOF</u>: As in the proof of theorem 1, $V[T_0,F_0] \leq V[T,F_0]$ for any regular translation invariant estimate T. (Note that $V[T_0^M,F_0] = V[T_0^{SA},F_0]$.) Taking $\theta=0$, it follows from the hypothesis on $m_F'(\theta)$ that

$$\int\psi_0'dF_0 = -\int\psi_0 f_0'dx = \int\psi_0^2 dF_0 = v_0,$$

and from this equation that

$$2m_F'(0) - v_0 = \int(2\psi_0' - \psi_0^2)d(F - F_0) + v_F.$$

Substituting the preceding equation into (5) gives

$$V[T_0^{SA},F_0] - V[T_0^{SA},F] = \frac{\int(2\psi_0'-\psi_0^2)d(F-F_0)}{v_0[2m_F'(0)-v_0]} \quad . \tag{10}$$

The hypothesis that $V[T_0^{SA},F]$ is positive and finite implies that the denominator in (10) is positive and hence $V[T_0,F_0] - V[T_0^{SA},F] \geq 0$ if and only if $\int(2\psi_0'-\psi_0^2)d(F-F_0) \geq 0$ which proves the part of the theorem regarding SA-estimates. It follows directly from (4) and (5) that

$$V[T_0^{SA},F] - V[T_0^{M},F] = \frac{V_F[m_F'(\theta)-v_0]^2}{v_0[2m_F'(\theta)-v_0][m_F'(\theta)]^2} .$$

The hypothesis that both $V[T_0^{SA},F]$ and $V[T_0^{M},F]$ are positive and finite implies that (11) is non-negative, and zero if and only if $m_F'(\theta) = v_0$. The rest of the proof now follows from the result just proved for SA-estimates.

While theorem 2 is appealing as a theoretical result, it would be of little practical value without some way to verify that the required inequality (9) holds other than checking each pdf in C individually. The next theorem gives conditions which are easily checked in most cases, and which guarantee that the inequality (9) holds over all of C. Thus it provides a practical way to utilize the result of theorem 2.

REMARK: From equation (11) one obtains that $V[T_0^{M},F] \leq V[T_0^{SA},F]$, with equality if and only if $m_F'(\theta) = v_0$. Thus, in general, the minmax M-estimate converges faster then the corresponding SA-estimate.

In the theorem that follows, we use the same symbol F to denote both a pdf and the corresponding measure on the Borel sets of $\overline{R}$, the extended real line.

THEOREM 3: Assume that (T_0,F_0) solves Problem 1 over C^* and that the following conditions hold:

(i) $T_0^{M}(T_0^{SA})$ is consistent and asymptotically normal with positive finite variance given by (4)((5)).

(ii) ψ_0 is absolutely continuous (but not necessarily bounded) on $\overline{R}$.

(iii) There is a subset A of $\overline{R}$ such that $F(A) = 0$ for each F in C and ψ_0' is bounded and upper semi-continuous on $\overline{R}$-A.

(iv) For each F in C there is a sequence $\{F_n\}$ from C^* such that $\int\phi dF_n \to \int\phi dF$ for any bounded and continuous ϕ.

Then (T_0,F_0) solves Problem 1 over C.

PROOF: First, we show that $m_F'(0) = \int\psi_0' dF$ $(\theta = 0)$. For each $t \neq 0$ let

$$\phi_t(x) = \min\{M, \max(-M, [\psi_0(x) - \psi_0(x-t)]/t)\},$$

where M is a bound for $|\psi_0'|$ on $\bar{R}$-A. Then $|\phi_t| \leq M$, ϕ_t is F integrable and $\phi_t \to \psi_0'$ F a.e. as $t \to 0$ for each F in C. By Lebesgue's dominated convergence theorem,

$$m_F'(0) = \frac{d}{dt}\int -\psi_0(x-t)dF(x)\Big|_{t=0} = \lim_{t\to 0}\int\phi_t dF = \int\psi_0' dF$$

for each F in C.

To complete the proof we show that the inequality (9) holds over C and the result follows from theorem 2.

Since (T_0, F_0) is a solution over C^* by hypothesis, theorem 2 implies that (9) holds over C^*. Let B be the set of all closure points of $\bar{R}$ - A which are in A. For each x in B, let $B(i,x) = \{y: |y - x| < 1/i \text{ and } y \in \bar{R} - A\}$. Define the function γ by the formula

$$\gamma = \begin{cases} 2\psi_0' - \psi_0^2 & \text{on } \bar{R} - A \\ \lim\limits_{i\to\infty} \sup\limits_{B(i,.)} (2\psi_0' - \psi_0^2) & \text{on } B \\ 2M - \psi_0^2 & \text{on A-B.} \end{cases}$$

Since ψ_0' is upper semi-continuous (usc) and bounded on $\bar{R}$-A and ψ_0 is continuous, it follows that γ is usc and bounded above. Let $\{\gamma_m\}$ be a sequence of bounded continuous functions decreasing pointwise to γ (see Ash [1], p. 390, for proof of existence). Let F be in C and let $\{F_n\}$ be a sequence from C^* with the convergence property of condition (iv). By the monotone convergence theorem

$\int\gamma_m dF \downarrow \int(2\psi_0' - \psi_0^2)dF$. Now for each m,

$$\limsup_n \int(2\psi_0' - \psi_0^2)dF_n \leq \limsup_n \int\gamma_m dF_n = \int\gamma_m dF.$$

Taking the limit as $m \to \infty$ we get

$$\limsup_n \int(2\psi_0' - \psi_0^2)dF_n \leq \int(2\psi_0' - \psi_0^2)dF.$$

Since (9) holds for each F_n the preceding inequality implies that it holds also for F.

The conditions of theorem 3 will typically be easy to check once the pdf F_0 is known. The available sufficient conditions for asymptotic normality are used to check (i). With F_0 known, a formula for ψ_0 is in hand. Thus (ii) is easily checked. The set A in condition (iii) is usually the set of points at which ψ_0' does not exist. If this is the case, condition (iii) is satisfied if all pdf's placing positive mass on A are excluded from C. As an example, take the p-point class of Martin and Masreliez [9]. Here, for a given set in the class, ψ_0' does not exist at two points $(\pm y_p)$. Distributions placing positive mass at these point are excluded from the set. Condition (iv) is probably the most difficult to verify.

I.3 DENSENESS OF DISTRIBUTIONS OF FINITE FISHER INFORMATION

Our next objective is to show that condition (iv) of theorem 3 holds on distribution sets in three general classes. Two of these, the ε - G and ε-contaminated G classes, were introduced by Huber [2] and have been considered by other authors in various robustness studies [6,7,8,9,13]. The third is a class which we call the generalized moment constrained (GMC) class. This class includes as special cases the p-point class considered by Martin and Masreliez [9] and the class of sets consisting of all pdf's taking common values at a specified finite set of points considered by Huber [5]. The approach is topological. We give the set P of all pdf's on the extended real line $\bar{R}$ the weakest topology under which all the maps $F \to \int \phi dF$ are continuous, where ϕ is bounded and continuous. This corresponds to the weak* topology induced on P as a subset of NBV[$\bar{R}$], the normalized space of functions of bounded variation on $\bar{R}$ [14]. In all subsequent statements which mention topological properties (such as closure, continuity, denseness, etc.) associated with distribution sets, this is the topology which is implied.

Condition (iv) of theorem 3 is equivalent to the requirement that C^* be dense in C. It is shown in [10] that the pdf's of finite Fisher information are dense in P. If C has non-empty interior relative to P and is contained in the closure of its interior, then it follows that C^* is dense in C. This is the approach we take to establish that condition (iv) holds.

The first considered is the ε-contaminated G class. Recall that the distribution set C is from the ε-contaminated G class if $C = \{F \in P : F = (1-\varepsilon) G + \varepsilon H$ for some H in P$\}$ for some fixed positive ε less than one.

<u>LEMMA 1</u>: If C is in the ε-contaminated G class and G has finite Fisher information, then C^* is dense in C.

Next we consider the ε-G class. For convenient reference, we note that if $0<\varepsilon<1$, and G is some pdf, then the set C is in the ε-G class if $C = \{F \in P : |F - G| \leq \varepsilon \text{ on } \overline{R}\}$. The condition given for guaranteeing denseness of C^* in the ε-G set C is less restrictive than the corresponding condition of lemma 1 for an ε-contaminated G set.

<u>LEMMA 2</u>: Suppose C is in the ε-G class, where B is a continuous pdf. Then C^* is dense in C.

The result of lemma 2 does not hold for arbitrary G. For example, if G has a discontinuity of magnitude $\delta>0$ at x_0 and $\varepsilon<\delta/2$, then each pdf in the ε-G set is discontinuous at x_0 and hence has infinite Fisher information. C^* is empty in this case.

Now we show that C^* is dense in C for sets of the type

$$C = \{F \in P : \int s_n dF \leq P_n,\ 1 \leq n \leq N\} . \tag{12}$$

This class includes cases in which the distribution set is a general GMC set

$$C_a = \{F \in P : \int s_n dF = P_n,\ 1 \leq n \leq M, \text{ and } \int s_n dF \leq P_n,\ M+1 \leq N \leq N\}$$

in the sense that if F_0 has minimum Fisher information in C_a, then there exist constants $q_n = \pm 1$, $1 \leq n \leq M$, such that F_0 also minimizes Fisher information over

$$C_b = \{F \in P : q_n \int s_n dF \leq q_n p_n,\ 1 \leq n \leq M, \text{ and } \int s_n dF \leq p_n,\ M+1 \leq n \leq N\} . \tag{13}$$

Details are given in [10].

A set of this form conveniently models the error distribution when available information is in the form of moments, partial moments, or a histogram. This general class includes the p-point family of Martin and Masreliez [9] as well as the sets consisting of al pdf's having common values at a finite number of specified points considered by Huber [5].

<u>LEMMA 3</u>: Suppose C is defined by (12) and s_n is upper semi-continuous on $\overline{R}$ and bounded above on finite intervals, $1 \leq n \leq N$. If for some F_1 in C and for each n, $\int s_n dF_1 < p_n$, then C^* is dense in C.

Example:

$$C = \{F\varepsilon\Omega: \quad dF = 1, \int_{-1}^{1} dF = p_2, \int_{-1}^{1} x^2 \, dF \leq p_3\} , \qquad (14)$$

Requiring that a solution be symmetric, positive on R, zero at infinity, and have a continuous derivative at ± 1 gives the formula

$$f_0(x) = \begin{cases} c_1 \sum_{n=0}^{\infty} \left(\sum_{k=0}^{n} A_{2k} A_{2(n-k)} \right) x^{2n}, & |x| \leq 1 \\ c_2 \exp\left(-\lambda_1^{.5}(|x| - 1)\right), & |x| \geq 1, \end{cases} \qquad (15)$$

where $A_0 = 1$, $A_2 = (\lambda_1 + \lambda_2)/8$, and

$$A_{2(n+1)} = \frac{(\lambda_1 + \lambda_2) A_{2n} + \lambda_3 A_{2(n-1)}}{8(n+1)(2n+1)} , \quad n \geq 1.$$

The parameters c_1, c_2, λ_1, λ_2, and λ_3 are determined from (15) by enforcing the requirements that f_0' be continuous and F_0 be in C.

The variety of solutions for different cases of this example is interesting. For a fixed value of p_2, let G be the solution when p_3 is so large that G satisfies the third constraint defining C with strict inequality. Put $\gamma = \int x^2 s_2 dG$. If $p_3 \geq \gamma$, then the distribution F_0 of minimum Fisher information in C is G. In this case F_0 is one of the p-point solutions of Martin and Mazreliez [9]. There is a value of p_3 slightly smaller than γ for which F_0 is Gaussian in form on the interval (-1,1). F_0 corresponds to the ε-contaminated normal solution of Huber [2] for the appropriate value of ε in this case. If p_3 is decreased further, F_0 gives rise to a redescending ψ_0 function. Figure (1) shows solutions for several different cases which demonstrate the properties just discussed.

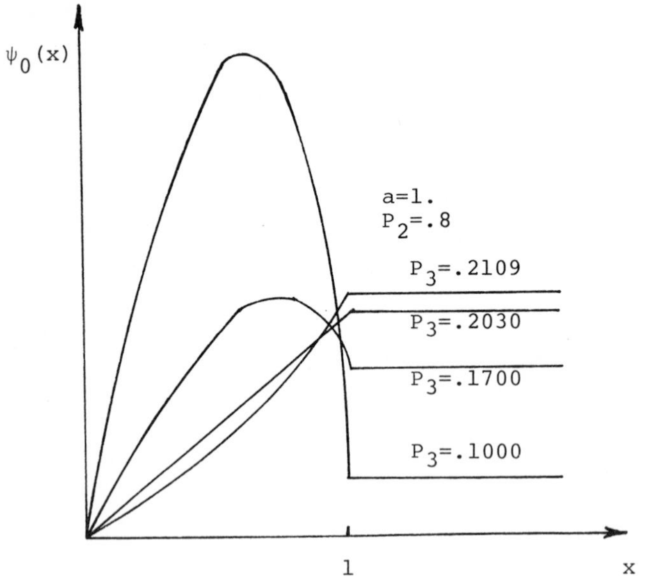

Figure 1. ψ_0 for solutions of the example

II. DETECTION

II.1 FIXED SAMPLE SIZE PROBLEM

Consider the following binary hypothesis testing problem:

$$H_0: \quad x_i = w_i$$

$$H_1: \quad x_i = \theta_1 s_i + w_i, \; i = 1,\ldots, N, \tag{16}$$

where $\{x_i\}$ is the sequence of observations, $\theta_1 > 0$, $\{s_i\}$ is a sequence of known constants with $|s_i| < \infty$ and

$$\lim_{N\to\infty} \sum_{i=1}^{N} s_i^2/N = c \tag{17}$$

and $\{w_i\}$ is a sequence of independent identically distributed random variables with a density, f, a member of a given class of symmetric densities, F. We are interested in finding a decision rule for detection that will provide guaranteed asymptotic performance - no matter which density function is picked from F.

Thus, we shall consider the following maxmin detection problem:

$$\max_{d\varepsilon D} \min_{f\varepsilon F} \beta_d(\theta_1|f) \text{ with } \beta_d(0|f) \leq \alpha \quad \forall\ f\varepsilon F \tag{18}$$

where $\beta_d(0|f)$ and $\beta_d(\theta_1|f)$ are the probability of false alarm and detection respectively when the true density function is $f(\cdot)$ and using detector d. Before solving the problem, a few additional definitions are necessary. Define a class of functions C, such that $L \varepsilon C$ if

1) L is convex, symmetric about the origin and strictly increasing for positive argument.
2) $\ell(t) = dL(t)/dt$ is continuous for all t.
3) For all $f \varepsilon F$, $E_f[\ell^2(x)] < \infty$.
4) For all $f \varepsilon F$, $\partial E_f[\ell(x-\theta)]/\partial\theta$ exists and is not equal to zero in some neighborhood of the origin.

Also, define θ_N as the value of θ which minimizes

$$\sum_{i=1}^{N} L(x_i-\theta s_i). \tag{19}$$

Note that θ_N in this definition is a generalized version of the M-estimate of location as defined by Huber [2].

The main results of this section will be obtained by verifying the following two inequalities

$$\beta_d(\nu|f_0) \leq \beta_{L_0}(\nu|f_0) \quad \forall\ d\varepsilon D \tag{20}$$

and

$$\beta_{L_0}(\nu|f_0) \leq \beta_{L_0}(\nu|f) \quad \forall \quad f \varepsilon F, \tag{21}$$

where $\beta_d(\nu|f) = \lim_{N\to\infty} \beta_d(\theta_1|f)$ with $\theta_1 = \nu/\sqrt{N}$ and both inequalities are subject to the constraint (18) on the false alarm rate. The subscript L_0 refers to a detector based on a threshold test using the generalized M-estimate derived from the function $L_0 \varepsilon C$ as the test statistic. Note that verification of (20) and (21) provides a slightly stronger result than called for by (18), in that their satisfaction makes (L_0, f_0) a saddlepoint pair. Assume, for the moment, that F contains a density of minimum Fisher information for location. Also, assume that the class of functions, C, contains a function L^* such that

$$L^*(x) = -\log f^*(x), \tag{22}$$

where f^* is the minimizing density defined above. For this case, we will show that $(L^*, f^*) = (L_0, f_0)$ is a saddlepoint pair. First, we require a preliminary lemma concerning the behavior of the generalized M-estimate.

LEMMA 4: Whenever $L \varepsilon C$ and $f \varepsilon F$, $\sqrt{N}(\theta_N - \theta_0)$ is asymptotically distributed as a zero-mean, normal random variable. The parameter θ_0 is the true value (0 under H_0 and θ_1 under H_1) of θ, and the variance is

$$V^2(f,L) = \frac{\int_{-\infty}^{\infty} \ell^2(x) f(x)\,dx}{c(\partial E_f[\ell(x-\theta)]/\partial\theta|\theta=0)^2} \quad . \tag{23}$$

From this result we conclude that a test (of size α) of the hypotheses based on θ_N is a consistent test, that is,

$$\lim_{N\to\infty} \beta_L(\theta_1|f) = 1 \tag{24}$$

for any fixed value of $\theta_1 \neq 0$. Also, the asymptotic power of a threshold test based on $\sqrt{N}\theta_N$ satisfies

$$\beta_L(\nu|f) = 1-\Phi\left(\frac{\gamma-\nu}{V(f,L)}\right). \tag{25}$$

where γ is the threshold of the test and Φ is the Gaussian distribution function with zero mean and unit variance.

LEMMA 5: If there exists a $f_0 \varepsilon F$ such that $L_0 = -\log f_0 \varepsilon C$ and if

$$V^2(f,L_0) \leq V^2(f_0,L_0) \quad \forall \ f\varepsilon F, \tag{26}$$

Then, the following relationships hold on the asymptotic power function

$$\sup_{L\varepsilon C} \beta_L(\nu|f_0) = \beta_{L_0}(\nu|f_0) = \inf_{f\varepsilon F} \beta_{L_0}(\nu|f), \tag{27}$$

whenever $\nu \geq \gamma$ and

$$\beta_{L_0}(0|f) \leq \beta_{L_0}(0|f_0) \quad \forall \ f\varepsilon F. \tag{28}$$

PROOF: By lemma 4, the test statistic is asymptotically distributed as a normal random variable for any $L\varepsilon C$. Thus, we have a test of the means $(0,\theta_1)$ of two normal random variables with the same variance. Equation (28) and the right hand equality of (27) follow from the monotonicity of the normal distribution function with respect to its variance. Also, since $L_0 = -\log f_0 \varepsilon C$, θ_N will be the maximum likelihood estimate of the mean of $f_0(x,\theta)$ which is efficient under the above conditions on C. Thus

$$V^2(f_0,L_0) = \frac{1}{cI(f_0)} \leq V^2(f_0,L) \quad \forall \quad L\varepsilon C \tag{29}$$

and the left hand equality of (27) follows by the same reasoning as above.

Lemma 4 shows that the pair (L_0,f_0) satisfies (18) and (21). To show that it also satisfies (20), we note that if $L_0 = -\log f_0$, then $\theta_N \equiv \hat{\theta}$, the maximum likelihood estimate if the true density function if f_0. Then, the following lemma gives the required result.

LEMMA 6: In testing the hypotheses described by (16), if the true density is f and $(-\log f) \ \varepsilon \ C$, then the test based on $\hat{\theta}$ (i.e., the solution to (19) with $L = -\log f$) is asymptotically equivalent to the likelihood ratio test, that is the test is asymptotically most powerful.

The lemma is a special case of a theorem by Wald [17] and the general result coincides with that by Chernoff [18]. The results of the above lemmas can be collected to give the following theorem.

THEOREM 4: In testing the hypothesis H_0 against the alternative H_1 described by (16), if there exists a $f_0 \varepsilon F$ such that $L_0 = -\log f_0 \ \varepsilon \ C$ and if

$$V^2(f,L_0) \leq V^2(f_0,L_0) \quad \forall \ f \varepsilon F, \tag{30}$$

then the following relations hold when ν is greater than or equal to γ, the threshold.

$$\beta_{L_0}(0|f) \leq \beta_{L_0}(0|f_0) \quad \forall \ f \varepsilon F \tag{31}$$

$$\inf_{f \varepsilon F} \beta_{L_0}(\nu|f) = \beta_{L_0}(\nu|f_0) = \sup_{d \varepsilon D} \beta_d(\nu|f_0) \tag{32}$$

If ν is less than γ, equations (20) and (21) cannot simultaneously be satisfied. The situation is similar to that discussed in Section I. The difficulty can be alleviated by increasing the number of observations taken. If this is not possible, the appropriate robust detector is one maximizing the minimum slop of the power function at the origin. Such a detector has been proposed in [6] and [7] for the special case of contaminated distributions. The following lemma shows that the detector proposed in this paper satisfies this criterion in a more general context.

LEMMA 6: Under the same assumptions on f_0 required in the previous theorem, a threshold test based on θ_N (derived from L_0) asymptotically maximizes the minimum slope of the power function at the origin with $\hat{f} = f_0$.

PROOF: By choosing γ to satisfy the false alarm requirement exactly for the least favorable density, f_0, we can rewrite (25) as

$$\beta_{L_0}(\nu|f) = 1-\Phi(r\Phi^{-1}(1-\alpha)-\nu/V(f,L_0)), \tag{33}$$

where $r = V(f_0,L_0)/V(f,L_0) \geq 1$. Taking the ratio of the partial derivatives evaluated at $\nu=0$ gives

$$\frac{\beta'_{L_0}(0|f_0)}{\beta'_{L_0}(0|f)} = \exp(-\tfrac{1}{2}[\phi^{-1}(1-\alpha)]^2(1-r^2))/r \tag{34}$$

This expression does not exceed one as long as

$$\alpha \geq 1-\phi([\log(r^2)/(r^2-1)]^{0.5})$$

The above results show that the design of the receiver will depend on finding the least favorable density function f_0 as defined in lemma 5. To find this density, we notice that it must have a finite Fisher information number for location according to assumption 3 on C. Thus, it will also be least favorable in the family of density functions, G, defined as

$$G = \{f:\ f\varepsilon F \text{ and } I(f) < \infty\}. \tag{35}$$

However, the least favorable density from this class is the one with minimum Fisher information (Theorem 2 [4]). Thus, to find f_0 we follow the steps described below:

a) Find f^* such that $I(f^*) \leq I(f)\ \forall\ f\varepsilon G$. This can usually be done using Lagrange multiplier techniques.
b) Check that $-\log f^* \ \varepsilon\ C$.
c) Check that f^* maximizes $V^2(L^*,f)$ over F. In general if f^* does not maximize $V^2(L^*,f)\ \forall\ f\varepsilon F$, then f_0 will not exist, since in this case we have only two alternatives. Either there is no f which maximizes the variance over F or there exists a f which maximizes the variance but not in G, so $L = -\log f$ is not an element of C.

In summary, the results given above show that the most robust detector may be designed by finding f_0 maximizing

$$V^2(f,-\log f) = (cI(f))^{-1}\ \forall\ f\varepsilon F \tag{36}$$

and then using $L_0 = -\log (f_0)$ to define θ_N (if $L_0 \varepsilon C$) and basing the threshold detector on θ_N.

Notice that, under the conditions of the theorem, the results may also be used to solve detection problems for simple hypotheses with a minimum probability of error criterion, M-ary hypotheses problems, and one or two sided composite hypotheses problems. Also, because of the monotonicity of ℓ we need not actually find the M-estimate, θ_N. It is sufficient to perform the following test

$$\sum_{i=1}^{N} \ell(x_i-\gamma) \lessgtr 0 \tag{37}$$

where γ is the threshold in the original test.

References [19] and [20] contain examples and more details about the above detection procedures.

II.2 SEQUENTIAL DETECTION

In some cases, we must consider detection when the observations are costly or when time is a major consideration. For these instances a robust sequential detector may be most appropriate. We will be interested in finding a sequential method for testing the hypothesis H_0 against the hypothesis H_1 defined as

$$\begin{aligned} H_0&: \quad x_i = \theta_0 + w_i \\ H_1&: \quad x_i = \theta_1 + w_i \end{aligned} \qquad i = 1,\ldots, N \tag{38}$$

where $\{x_i\}$ is the sequence of observations, θ_i is the signal under H_i and $\{w_i\}$ is a sequence of independent identically distributed noise components with density function, f, which is an unknown member of a class of symmetric density functions, F. As in the previous section, under some assumptions on the class F, a threshold test of hypotheses based on the M-estimate of the mean of the x_i's, will be an asymptotically most robust test of the above hypotheses in a maximum sense. Here we treat the same problem imposing additional restrictions on the test by assuming that $|\theta_1-\theta_0| = \delta > 0$ where δ is a very small number and we want to design a sequential test such that the probability of false alarm $P_F \leq \alpha$ and the probability of missing $P_M \leq \beta$ for all $f \varepsilon F$, where α and β are some small positive numbers. Although the method proposed before is still applicable to this problem, the required number of observations may be very large, as is the case with other threshold tests, and consequently the time for processing will increase and the need arises for a practical procedure to decrease the required number of observations.

In the following we provide a brief description of two strategies, called M-sequential (MS) detector and stochastic approximation sequential (SAS) detector, for the design of sequential robust detectors. More details about these methods, as well as, a comparison between them can be found in [20].

Assume that there exists a density function $f_0 \varepsilon F$ such that $L_0 = -\log f_0 \varepsilon C$ and

$$\sup_{f\varepsilon F} V^2(L_0,f) = V^2(L_0,f_0) = V_0{}^2 = \inf_{L\varepsilon C} V^2(L,f_0). \qquad (39)$$

The MS detector will be defined as:

Accept H_0 if

$$T_N = \frac{N(\theta_1-\theta_0)}{V_0{}^2}\{\theta_N-\tfrac{1}{2}(\theta_1+\theta_0)\} \leq A\ , \qquad (40)$$

Reject H_0 if

$$T_N \geq B\ . \qquad (41)$$

Continue taking observations if

$$A < T_N < B. \qquad (42)$$

Here A and B are two constants chosen to make the probability of false alarm $P_F=\alpha$ and the probability of missing $P_M = (1-P_D) = \beta$ when the noise density function is f_0 and θ_N is computed using $L=L_0$.

The following results give the main properties of the MS-detector as defined above.

<u>LEMMA 7</u>: Suppose A and B are finite, $\theta_1 \neq \theta_0$ and $V^2(L_0,f) \neq 0$ for all density functions $f\varepsilon F$. Then for any $f\varepsilon F$, the MS-test will terminate with probability one at a finite number of observations, that is

$$P_{\theta i}[N<\infty] = 1 \qquad (43)$$

Assume that there is some parameter Q, such that as $Q\to\infty$ the relevant sample size is proportional to Q and $\lim_{Q\to\infty} \sqrt{Q}\,|\theta_1-\theta_0| = k_Q$ and $\lim_{Q\to\infty} [N_i(f)/Q] = E_i(N\ f)$, where $N_i(f)$ is the expected number of observations required for the MS-test to terminate when the noise density function is f and the true hypothesis is H_i. Also, define the risk function R(d,f), when the detection strategy d of size (α,β) is used and the noise density function is f, as

$$R(d,f) = r_0\alpha + r_1\beta + c_0\ E_0(N|d,f) + c_1\ E_1(N|d,f) \tag{44}$$

where (r_0,r_1,c_0,c_1) is a set of costs, $E_i(N|d,f) = \lim_{Q\to\infty} N_i(d,f)/Q$, and $N_i(d,f)$ is the expected sample size under H_i. Then we have the following theorem.

<u>THEOREM 5</u>: In testing the hypothesis H_0 against the hypothesis H_1 given by equation (38) the MS-detector will satisfy the following relations asymptotically (i.e. as $Q\to\infty$)

1) $$\begin{aligned} P_F(f) &\le P_F(f_0) \le \alpha \\ P_M(f) &\le P_M(f_0) \le \beta \end{aligned} \qquad \text{for all } f\varepsilon F \tag{45}$$

2) $$E_0(N|f) < \frac{-2V_0{}^2A}{K_Q{}^2} = E_0{}^* \quad \text{for all } f\varepsilon F \tag{46}$$

$$E_1(n|f) < \frac{2V_0{}^2B}{K_Q{}^2} = \Xi_1{}^* \quad \text{for all } f\varepsilon F \tag{47}$$

3) There exist a set of costs such that

$$\inf_{d\varepsilon D} R(d,f_0) = R(L_0,f_0) = \sup_{f\varepsilon F} R(L_0,f) \tag{48}$$

where $R(L_0,.)$ is the risk function for the MS-detector and D is the class of all detectors.

The SAS-detector has essentially the same construction as the MS-detector except the M-estimate of location θ_N is replaced by the stochastic approximation estimate S_N.

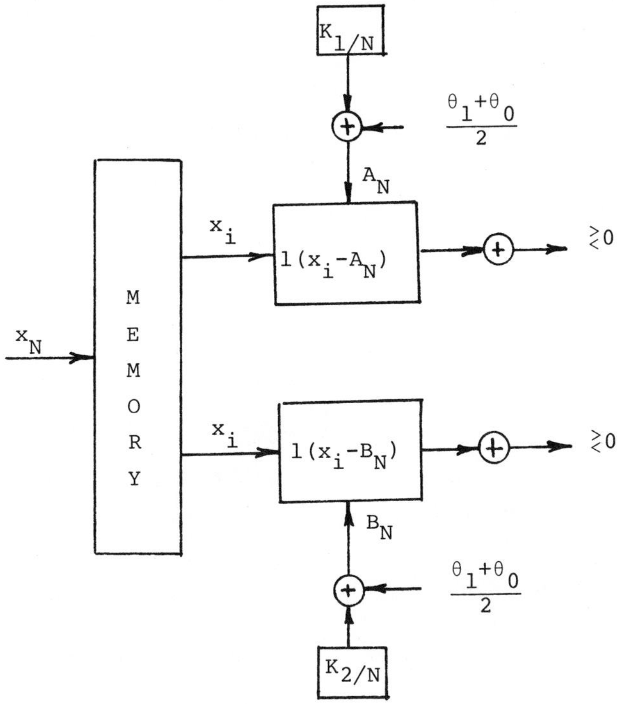

Figure 2. MS-detector implementation

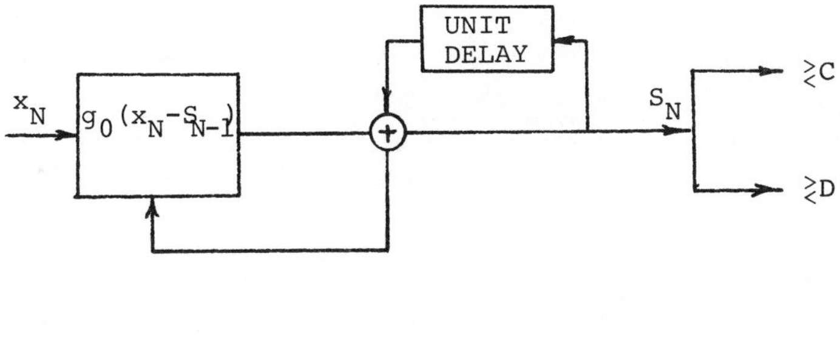

Figure 3. SAS-detector implementation

It has the same properties as described above for the MS-estimate. A simple implementation of both detectors is shown in Figures (2) and (3) where

$$k_1 = \frac{V_0^2 A}{(\theta_1 - \theta_0)} \quad \text{and} \quad k_2 = \frac{V_0^2 B}{(\theta_1 - \theta_0)} \quad .$$

REFERENCES

1. Ash, R. B. (1972). Real Analysis and Probability, Academic Press, New York.
2. Huber, P. J. (1964). Robust Estimation of a Location Parameter, Ann. Math. Statist., vol. 35, pp. 73-101.
3. Huber, P. J. (1967). The Behavior of Maximum Likelihood Estimates Under Nonstandard Conditions, Proc. Fifth Berkeley Symp. Math. Statist., vol. 1, pp. 221-233.
4. Huber, P. J. (1972). The 1972 Wald Lecture, Robust Statistics: A Review, Ann. Math. Statist., vol. 43, pp. 1041-1067.
5. Huber, P. J. (1974). Fisher Information and Spline Interpolation, Ann. Statist., vol. 2, pp. 1029-1033.
6. Kassam, S. A. and Thomas, J. B. (1976). Asymptotically Robust Detection of a Known Signal in Contaminated Non-Gaussian Noise, IEEE Trans. Inform. Theory, vol. IT-22, pp. 22-26.
7. Martin, R. D. and Schwartz, S. C. (1971). Robust Detection of a Known Signal in Nearly Gaussian Noise, IEEE Trans. Inform. Theory, vol. IT-17, pp. 50-56.
8. Martin, R. D. (1972). Robust Estimation of Signal Amplitude, IEEE Trans. Inform. Theory, vol. IT-18, pp. 596-606.

9. Martin, R. D. and Masreliez, C. J. (1975). Robust Estimation via Stochastic Approximation, IEEE Trans. Inform. Theory, vol. IT-21, pp. 263-271.

10. Price, E. L. (1977). Robust Estimation by Stochastic Approximation and Error Models for Robust Procedures, Ph.D. Dissertation, The Johns Hopkins University, Baltimore, MD.

11. Prokhorov, Y. V. (1956). Convergence of Random Processes and Limit Theorems in Probability Theory, Theory of Prob. and its Appl., vol. 1, pp. 157-214.

12. Sacks, J. (1958). Asymptotic Distribution of Stochastic Approximation Procedures, Ann. Math. Statist., vol. 29, pp. 373-405.

13. Sacks, J. and Ylvisaker, D. (1972). A Note on Huber's Robust Estimation of a Location Parameter, Ann. Math. Statist., vol. 43, pp. 1068-1075.

14. Taylor, A. E. (1958). Introduction to Functional Analysis, Wiley, New York.

15. Tukey, J. W. and Harris, T. E. (1949). Statistical Res. Group, Princeton Univ., Princeton, NJ, Memo. Rep. 31.

16. Tukey, J. W. (1960). A Survey of Sampling from Contaminated Distributions, in Contributions to Probability and Statistics, (Harold Hotelling Volume), Stanford, California: Stanford Univ. Press, p. 448.

17. Wald, A. (1941). Asymptotically most Powerful tests of Statistical Hypotheses, Ann. Math. Statist., Vol. 12, pp. 1-19.

18. Chernoff, H. (1954). On the Distribution of the Likelihood ratio, Ann. Math. Statist., Vol. 25, No. 3, pp. 573-578.

19. El-Sawy, A. H. and VandeLinde, V. D. (1977). Robust Detection of Known Signals, IEEE Trans. Inform. Theory, Vol. IT-23, No. 6, pp. 722-727.

20. El-Sawy, A. H. (1977). Robust Detection of Signals in Noise, Ph.D. Dissertation, The Johns Hopkins University, Baltimore, MD.

Department of Electrical Engineering
Johns Hopkins University
Baltimore, Maryland 21218

Robustness in the Strategy of Scientific Model Building

G. E. P. Box

Robustness may be defined as the property of a procedure which renders the answers it gives insensitive to departures, of a kind which occur in practice, from ideal assumptions. Since assumptions imply some kind of scientific model, I believe that it is necessary to look at the process of scientific modelling itself to understand the nature of and the need for robust procedures. Against such a view it might be urged that some useful robust procedures have been derived empirically without an explicitly stated model. However, an empirical procedure implies some unstated model and there is often great virtue in bringing into the open* the kind of assumptions that lead to useful methods. The need for robust methods seems to be intimately mixed up with the need for simple models. This we now discuss.

*An example (1), (2) was the application in the 1950's of exponential smoothing for business forecasting and the wide adoption in this century of three-term controllers for process control. It was later realized that these essentially empirical procedures point to the usefulness of ARIMA time series models since both are optimal for disturbances generated by such models.

ISBN 0-12-438150-2

THE NEED FOR SIMPLE SCIENTIFIC MODELS - PARSIMONY

The scientist, studying some physical or biological system and confronted with numerous data, typically seeks for a model in terms of which the underlying characteristics of the system may be expressed simply.

For example, he might consider a model of the form

$$y_u = f^{(p)}(\underset{\sim}{\xi}_u, \underset{\sim}{\theta}) + \varepsilon_u \qquad (u = 1,2,\ldots,n) \tag{1}$$

in which the expected value η_u of a measured output y_u is represented as some function of k inputs $\underset{\sim}{\xi}$ and of p parameters $\underset{\sim}{\theta}$, and ε_u is an "error". One important measure of simplicity of such a model is the number of parameters that it contains. When this number is small we say the model is parsimonious.

Parsimony is desirable because (i) when important aspects of the truth are simple, simplicity illuminates, and complication obscures; (ii) parsimony is typically rewarded by increased precision (see Appendix 1); (iii) indiscriminate model elaboration is in any case not a practical option because this road is endless*.

ALL MODELS ARE WRONG BUT SOME ARE USEFUL

Now it would be very remarkable if any system existing in the real world could be exactly represented by any simple model. However, cunningly chosen parsimonious models often do

---

*Suppose for example that in advance of any data we postulated a model of the form of (1) with the usual normal assumptions. Then it might be objected that the distribution of ε_u might turn out to be heavy-tailed. In principle this difficulty could be allowed for by replacing the normal distribution by a suitable family of distributions showing varying degrees of kurtosis. But now it might be objected that the distribution might be skew. Again, at the expense of further parameters to be estimated, we could again elaborate the class of distribution considered. But now the possibility might be raised that the errors could be serially correlated. We might attempt to deal with this employing, say, a first order autoregressive error model. However, it could then be argued that it should be second order or that a model of some other type ought to be employed. Obviously these possibilities are extensive, but they are not the only ones: the adequacy of the form of the function $f(\underset{\sim}{\xi}, \underset{\sim}{\theta})$ could be called into question and elaborated in endless ways; the choice of input variables $\underset{\sim}{\xi}$ might be doubted and so on.

provide remarkably useful approximations. For example, the law $PV = RT$ relating pressure P, volume V and temperature T of an "ideal" gas via a constant R is not exactly true for any real gas, but it frequently provides a useful approximation and furthermore its structure is informative since it springs from a physical view of the behavior of gas molecules.

For such a model there is no need to ask the question "Is the model true?". If "truth" is to be the "whole truth" the answer must be "No". The only question of interest is "Is the model illuminating and useful?".

ITERATIVE PROCESS OF MODEL BUILDING

How then is the model builder to know what aspects to include and what to omit so that parsimonious models that are illuminating and useful result from the model building process? We have seen that it is fruitless to attempt to allow for all contingencies in advance so in practice model building must be accomplished by iteration* the inferential stage of which is illustrated in Figure 1.

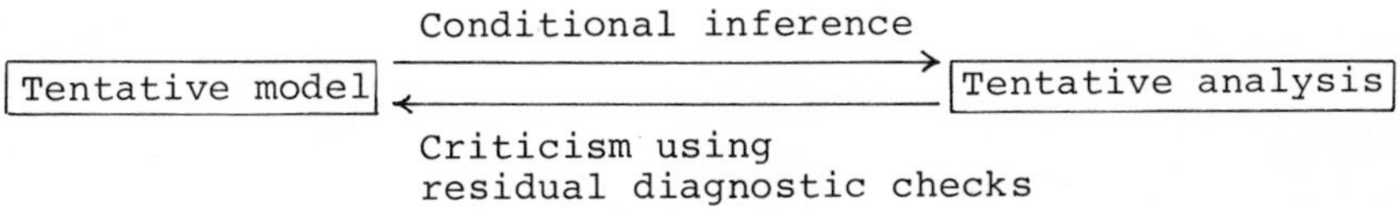

Figure 1. Interative Model Building

For example, preliminary graphical analysis of data, and careful thought about what is known about the phenomenon under

*The iterative building process for scientific models can take place over short or long periods of time, and can involve one investigator or many. One interesting example is the process of discovery of the structure of DNA described by J. D. Watson [3]. Another is the development by R. A. Fisher [4] of the theory of experimental design between 1922 and 1926. The recognition that scientific model building is an iterative process goes back to such classical authors as to Aristotle, Grossteste and Bacon. The suggestion that statistical procedures ought to be viewed in this iterative context was discussed for example in [1], [5], [6], [7], [22].

study, may suggest a first model worthy to be tentatively entertained. From this model a corresponding first tentative analysis may be made as if we believed it. The tentative inferences made, like all inferences are conditional on the applicability of the implied model, but the investigator now quickly switches his attitude from that of sponsor to that of critic. In fact the tentative analysis provides a basis for criticism of the model. This criticism phase is accomplished by examining residual quantities using graphical procedures and sometimes more formal tests of fit. Such diagnostic checks may fail to show cause for doubting the model's applicability, otherwise it may point to modification of the model leading to a new tentative model and, in turn, to a further cycle of the iteration.

WAYS TO ALLOW FOR MODEL DISCREPANCIES

How can we avoid the possibility that the parsimonious models we build by such an iteration might be misleading? There are two answers.

a) Knowing the scientific context of an investigation we can allow in advance for more important contingencies.
b) Suitable analysis of residuals can lead to our fixing up the model in other needed directions.

We call the first course model robustification the second iterative fixing.

JUDICIOUS MODEL ROBUSTIFICATION

Experience with data and known vulnerabilities of statistical procedures in a specific scientific context will alert the sensitive practitioner to likely discrepancies that can cause problems. He may then judiciously and grudgingly elaborate the model and hence the resulting procedure so as to insure against particular hazards in the most parsimonious manner. Models* providing for simple forms of

---

*It is currently fashionable to conduct robustness studies in which the normality assumption is relaxed (in favor of heavy tailed distribution or distributions containing outliers) but all other assumptions are retained. Thus it is still assumed that errors are independent that transformations are correctly specified and so on. This seems to be too naive and narrow a view.

autocorrelation in serial data, for simple transformation in data covering wide ranges, for outliers in the almost universal situation where perfect control of the experimental process is not available, are all examples of commonly needed parsimonious elaborations which can have major consequences.

ITERATIVE FIXING USING DIAGNOSTIC CHECKS

Once it is recognized that the choice of model is not an irrevocable decision, the investigator need not attempt to allow for all contingencies a priori which as we have said is in any case impossible. Instead, after appropriate robustification, he may look at residual quantities in an attempt to reveal discrepancies not already provided for.

To better appreciate such a process of iterative fixing, write the model (1) in the form

$$y_u = f(\underset{\sim}{\xi}_{1u}) + \varepsilon(\underset{\sim}{\xi}_{2u}) \qquad (2)$$

where now the vector $\underset{\sim}{\xi}_{1u}$ previously denoted in (1) by $\underset{\sim}{\xi}_{u}$ represents those variables the investigator has specifically decided to study. The expression $\varepsilon(\underset{\sim}{\xi}_{2u})$ which replaces ε_u indicates explicitly that the error ε_u represents the joint influences on the output of all those other input variables $\underset{\sim}{\xi}_{2u}$ which are omitted from the model (usually because they are unknown). Many statistical procedures (in particular quality control, residual analysis and evolutionary operation) are concerned with discovering "assignable causes" - elements of $\underset{\sim}{\xi}_{2u}$ - which may be moved out of the unknown to the known as indicated by

$$y_u = f(\overleftarrow{\underset{\sim}{\xi}_{1u}) + \varepsilon(\underset{\sim}{\xi}_{2u}}) \;. \qquad (3)$$

Now let $\{a_t\}$ be a white noise sequence. That is a sequence of identically and independently distributed random variables having zero mean. If we now denote the n values of response and known inputs by $\underset{\sim}{y}$ and $\underset{\sim}{\xi}_1$ respectively then an ideal model

$$F_t\{\underset{\sim}{y},\underset{\sim}{\xi}_1\} = a_t, \qquad t = 1,2,\ldots,n \qquad (4)$$

would consist of a transformation of the data to white noise which was statistically independent of any other input.

Iteration towards such a model is partially motivated by diagnostic checking through examination of residuals at each stage of the evolving model.

Thus, patterns of residuals can indicate the presence of outliers or suggest the necessity for including specific new inputs. Also serial correlation of residuals shown for example by plotting $\hat{a}_{t+k}$ versus $\hat{a}_t$ (or more formally by examining estimates of sample autocorrelations for low lags k) can point to the need for allowing for serial correlation in the original noise model. Again as was shown by Tukey [8], dependence of $y - \hat{y}$ on $\hat{y}^2$ can indicate the need for transformation of the output. Examination of residuals at each stage parallels the chemical examination of the filtrate from an extraction process. When we can no longer discover any information in the residuals then we can conclude that all extractable information is in the model.*

MODEL ROBUSTIFICATION AND DIAGNOSTIC CHECKING

Robustification and iterative fixing following diagnostic checking of residuals are of course not rival but complementary techniques and we must try to see how to use both wisely.

Subject matter knowledge will often suggest the need for robustification by parsimonious model elaboration. For example, when models such as (2) are used in economics and business the output and input variables $\{y_u\}$, $\{\underset{\sim}{\xi}_{1u}\}$ are often collected serially as <u>time series</u>. They are then very likely to be autocorrelated. If this is so then the

---

*It should be remembered that just as the Declaration of Independence promises the <u>pursuit</u> of happiness rather than happiness itself, so the iterative scientific model building process offers only the pursuit of the perfect model. For even when we feel we have carried the model building process to a conclusion some new initiative may make further improvement possible. Fortunately to be useful a model does not have to be perfect.

In particular notice that, even though residuals from some model are consistent with a white noise error, this does not bar further model improvement. For example, this white noise error could depend on (theoretically, even, be proportional to) the white noise component of some, so far unrecognized, input variable.

components of $\{\underset{\sim}{\xi}_{2u}\}$ in the error $\varepsilon\{\underset{\sim}{\xi}_{2u}\}$ are equally likely to have this characteristic. It makes little sense, in this context, therefore, to postulate even tentatively a model of the form of (1) in which the ε_u are supposed independent. Instead the representation of ε_u by a simple time series model (for example a first order autoregressive process for which serial correlation falls off exponentially with lag) would provide a much more plausible starting place. Failure either, to robustify the model in this way initially, or, to check for serial correlation in residuals, resulted in one published example in t values (measuring the significance of regression coefficients) which were inflated by an order of magnitude [9,10]. We discuss this example in more detail later.

Again statistical analysis in an inappropriate metric can lead to wasteful inefficiency. For instance, textile data are presented in [11] where appropriate transformation would have resulted in a three fold decrease in the relative variance accompanied by reduction in the number of needed parameters from ten to four, for the expenditure of only one estimated transformation parameter.

In both examples discussed above a profound improvement in statistical analysis is made possible by suitable robustification of the model the need for which could have been detected by suitable diagnostic checks on residuals.

AVOIDANCE OF UNDETECTED MISSPECIFICATION

Unfortunately we cannot always rely on diagnostic checks to reveal serious model misspecification. The dangerous situation is that where initial model misspecification can result in a seriously misleading analysis whose inappropriateness is unlikely to be detected by diagnostic checks.

For example, the widely used model formulation (1) supposes its applicability for every observation y_u $(u = 1,2,...,n)$ and so explicitly excludes the possibility of outliers. If such a model is (inappropriately) assumed in the common situations where occasional accidents possibly leading to outliers are to be expected, then any sensible method of estimation such as maximum likelihood if applied using this inappropriate model must tend to conceal model inadequacy. This is because, in order to follow the mathematical instructions presented, it must choose parameters which make residuals even with this wrong model look as much as possible like white noise. That this has led some investigators to abandon standard inferential methods rather than the misleading model seems perverse.

As a further example of the hazard of undetected misspecification consider scientific problems requiring the comparison of variances. Using standard normal assumptions the investigator might be led to conduct an analysis based on Bartlett's test. However this procedure is known to be so sensitive to kurtosis that nonnormality unlikely to be detected by diagnostic checks could seriously invalidate results. This characteristic of the test is well known, of course, and it has long been recognized that the wise researcher should robustify initially. That is he should use a robust alternative to Bartlett's test ab initio rather than relying on a test of nonnormality followed by possible fix up.

The conclusion is that the role of model robustification is to take care of likely discrepancies that have dangerous consequences and are difficult to detect by diagnostic checks. This implies an ability by statisticians to worry about the right things. Unfortunately they have not always demonstrated this talent, see for example Appendix 2.

ROBUSTNESS AND ERROR TRANSMISSION

Since we need parsimonious models but we know they must be false we are led to consider how much deviation from the model of a kind typically met in practice will affect the procedure derived on the assumption that a model is exact.

The problem is analogous to the classical problem of error transmission. In its simplest manifestation that problem can be expressed as follows:

Consider a calibration function

$$\gamma = f(\beta) \tag{5}$$

which is used to determine γ at some value say $\beta = \beta_0$. Suppose that the function is mistakenly evaluated at some other value of β, then the resulting error ε transmitted into γ is

$$(\gamma_0 + \varepsilon) - \gamma_0 = f(\beta) - f(\beta_0) \simeq (\beta - \beta_0) \times \rho \tag{6}$$

where $\rho = \left.\frac{\partial \gamma}{\partial \beta}\right|_{\beta=0}$.

The expression for the transmitted error ε contains *two* factors β and ρ. The first is the size of the input error the second which we will call the *specific transmission* is the rate of change of γ as β is changed. It is frequently emphasized in discussing error transmission that *both* factors are important. In particular the existence of a large discrepancy $\beta - \beta_0$ does not lead to a large transmitted error ε if ρ is small. Conversely even a small error β can produce a large error ε if ρ is large.

Now consider a distribution of errors $p(\beta)$. Knowledge of the relation $\gamma = f(\beta)$ allows us to deduce the corresponding distribution $p(\varepsilon)$. In particular if the approximation (6) may be employed then $\sigma_\gamma = p\sigma_\beta$. The relevance of the above robustness studies is as follows. Suppose γ is some performance characteristic of a statistical procedure which it is desired to study. This characteristic might be some measure of closeness of an estimate to the true value, significance level, the length of a confidence interval, a critical probability, a posterior probability distribution, or a rate of convergence of some measure of efficiency or optimality. Also suppose β is some measure of departure from assumption such as a measure of nonnormal kurtosis or skewness or autocorrelation of the error distribution and suppose that $\beta = \beta_0$ is the value taken on standard assumptions. Then in the error transmission problem three features of importance are

(1) The distribution of β. This provides the probability distribution of deviations from assumption which are actually encountered in the real world. Notice this feature has nothing to do with mathematical derivation or with the statistical procedure used.

(ii) The specific transmission ρ. This is concerned with the error transmission characteristics of the statistical procedure actually employed and may be studied mathematically. It is well known that different statistical procedures can have widely different ρ's. An example already quoted is the extreme sensitivity to distribution kurtosis of the significance level of likelihood ratio tests to compare variances (Bartlett's test) and the comparative insensitivity of corresponding tests to compare means (Analysis of variance tests).

(iii) If the data set is of sufficient size it can itself provide information about the discrepancy $\beta - \beta_0$ which occurs in that particular sample, thus reducing reliance on prior knowledge. Conversely if the sample size is small* or if β is of such a nature that a very large sample is needed to gain even an approximate idea of its value, heavier reliance must be placed on prior knowledge (whether explicitly admitted or not).

It seems to me that these three characteristics taken together determine what we should worry about. They are all incorporated precisely and appropriately in a Bayes formulation.

BAYES THEOREM AS A MEANS OF STUDYING ROBUSTNESS

From a Bayesian point of view given data $\underset{\sim}{y}$ all valid inferences about paramaters $\underset{\sim}{\theta}$ can be made from an appropriate posterior distribution $p(\underset{\sim}{\theta}|\underset{\sim}{y})$. To study the robustness of such inferences when discrepancies β from assumptions occur we can proceed as follows:

Consider a naive model relating data $\underset{\sim}{y}$ and parameters $\underset{\sim}{\theta}$. For example, it might assume that $p(\underset{\sim}{y}|\underset{\sim}{\theta})$ was a spherically normal density function, that $E(\underset{\sim}{y})$ was linear in the parameters $\underset{\sim}{\theta}$ and that before the data became available the desired state of ignorance about unknown parameters was expressed by suitable non-informative prior distributions leading to the standard analysis of variance and regression procedures. Suppose it was feared that certain discrepancies

*However even the small amount of information about β available from a small sample can be important. See for instance the analysis of Darwin's data which follows (Example 1).

from the model might occur (for example lack of independence, need for transformation, existence of outliers, non-normal kurtosis etc.). Two questions of interest are (A) how sensitive are inferences made about θ to these contemplated misspecifications of the model? (B) If necessary how may such inferences be made robust against such discrepancies as actually occur in practice?

QUESTION (A) SENSITIVITY

Suppose in all cases that discrepancies are parameterized by $\underset{\sim}{\beta}$. Also suppose the density function for $\underset{\sim}{y}$ given $\underset{\sim}{\theta}$ and $\underset{\sim}{\beta}$ is $p(\underset{\sim}{y}|\underset{\sim}{\theta},\underset{\sim}{\beta})$ and that $p(\underset{\sim}{\theta}|\underset{\sim}{\beta})$ is a non-informative prior for $\underset{\sim}{\theta}$ given $\underset{\sim}{\beta}$. Then comprehensive inferences about $\underset{\sim}{\theta}$ given $\underset{\sim}{\beta}$ and $\underset{\sim}{y}$ may be made in terms of the posterior distribution

$$p(\underset{\sim}{\theta}|\underset{\sim}{\beta},\underset{\sim}{y}) = k\, p(\underset{\sim}{y}|\underset{\sim}{\theta},\underset{\sim}{\beta})p(\underset{\sim}{\theta}|\underset{\sim}{\beta}) \qquad (7)$$

where k is a normalizing constant. Sensitivity of such inferences to changes in $\underset{\sim}{\beta}$ may therefore be judged by inspection of $p(\underset{\sim}{\theta}|\underset{\sim}{\beta},y)$ for various values of $\underset{\sim}{\beta}$.

QUESTION (B) ROBUSTIFICATION

Suppose now that we introduce a prior density $p(\underset{\sim}{\beta})$ which approximates the probability of occurrence of $\underset{\sim}{\beta}$ in the real world. Then we can obtain $p(\underset{\sim}{\beta}|\underset{\sim}{y})$ from $\int p(\underset{\sim}{\theta},\underset{\sim}{\beta}|\underset{\sim}{y})d\underset{\sim}{\theta}$. This is the posterior distribution of $\underset{\sim}{\beta}$ given the prior $p(\beta)$ and given the data. Then

$$p(\underset{\sim}{\theta}|\underset{\sim}{y}) = \int p(\underset{\sim}{\theta}|\underset{\sim}{\beta},\underset{\sim}{y})p(\underset{\sim}{\beta}|\underset{\sim}{y})d\underset{\sim}{\beta} \qquad (8)$$

from which (robust) inferences may be made about $\underset{\sim}{\theta}$ independently of $\underset{\sim}{\beta}$ as required.*

Inference are best made by considering the whole posterior distribution however if point estimates are needed they can of course be obtained by considering suitable features of the posterior distribution. For example the posterior mean

---

*More generally the density function for y will contain nuisance parameters $\underset{\sim}{\sigma}$ Equations (7) and (8) will then apply after these parameters have been eliminated by integration.

will minimize squared error loss. Other features of the posterior distribution will provide estimates for other loss functions (see for example [6], [12]).

It does seem to me that the inclusion of a prior distribution is essential in the formulation of robust problems. For example, the reason that robustifiers favour measures of location alternative to the sample average is surely because they have a prior belief that real error distributions may not be normal but may have heavy tails and/or may contain outliers. They evidence that belief covertly by the kind of methods and functions that they favour which place less weight on extreme observations. I think it healthier to bring such beliefs into the light of day where they can be critically examined, compared with reality, and, where necessary, changed. Some examples of this alternative approach are now given.

EXAMPLE 1 KURTOSIS AND THE PAIRED t TEST

This section follows the discussion by Box and Tiao [6], [13], [14] of Darwin's data quoted by Fisher on the differences in heights of 15 pairs of self and cross-fertilized plants. These differences are indicated by the dots in Figure 2. The curve labeled $\beta = 0$ in that diagram is a t distribution centered at the average $\bar{y} = 20.93$ with scale factor $s/\sqrt{n} = 9.75$. On standard normal assumptions it can be interpreted as a confidence distribution, a fiducial distribution or a posterior distribution of the mean difference θ. From the Bayesian view point the distribution can be written

$$p(\theta|\underset{\sim}{y}) = \text{const}\left\{1 + \frac{n(\theta - \bar{y})^2}{\nu s^2}\right\}^{-\frac{n}{2}} \tag{9}$$

and results from taking a non-informative prior distribution for the mean θ and the standard deviation σ. Alternatively we may write the distribution (9) in the form

$$p(\theta|\underset{\sim}{y}) = \text{const}[\Sigma(y - \theta)^2]^{-\frac{n}{2}} \tag{10}$$

and if

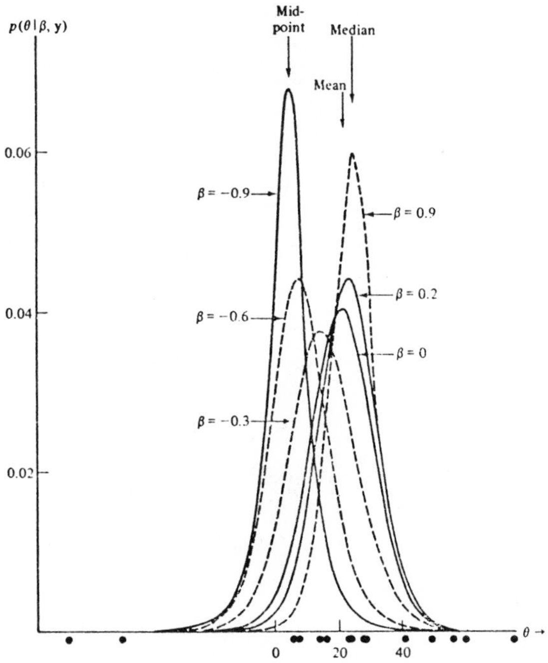

Figure 2. Posterior distributions $p(\theta|\beta,\underset{\sim}{y})$ of mean difference θ for parent distributions having differing amounts of kurtosis parameterized by β. Darwin's data.

$$M(\theta,q) = \Sigma|y_i - \theta|^q, \qquad q \geq 1$$

(10) may be written as

$$p(\theta|y) = \text{const}\{M(\theta,2)\}^{-\frac{n}{2}} . \tag{11}$$

SENSITIVITY TO KURTOSIS

One way to consider discrepancies arising from non-normal kurtosis is to extend the class of density functions, using the exponential power family

$$p(y|\theta,\sigma,\beta) = \sigma^{-1} \exp\left\{-c(\beta)\left|\frac{y - \theta}{\sigma}\right|^{2/(1+\beta)}\right\} \tag{12}$$

where with $c(\beta) = \left\{\frac{\Gamma[\frac{3}{2}(1 + \beta)]}{\Gamma[\frac{1}{2}(1 + \beta)]}\right\}^{\frac{1}{1+\beta}}$ and θ and σ are the mean and standard deviation as before. Then using the same noninformative prior distribution as before $p(\theta,\sigma|\beta) \propto \sigma^{-1}$ it is easily shown that in general

$$p(\theta|\beta,\underset{\sim}{y}) = \text{const } M\{\theta, 2/(1+\beta)\}^{-\frac{1}{2}n(1+\beta)} \tag{13}$$

and in particular if $\beta = 0$ (13) and (11) are identical.

The performance characteristic here is not a single quantity but the whole posterior distribution from which all inferences about θ can be made.

Sensitivity of the inference to changes in β is shown by the changes that occur in the posterior distributions $p(\theta|\beta,\underset{\sim}{y})$ when β is changed. Figure 2 shows these distributions for various values of β. Evidently, for this example, inferences* are quite sensitive to changes in the parent density involving more or less kurtosis.

ROBUSTIFICATION FOR KURTOSIS

As was earlier explained, high sensitivity alone does not necessarily produce lack of robustness. This depends also on how large are the discrepancies which are likely to occur, represented in (8) by the factor $p(\underset{\sim}{\beta}|\underset{\sim}{y})$. It is convenient to define a function $p_u(\underset{\sim}{\beta}|\underset{\sim}{y}) = p(\underset{\sim}{\beta}|\underset{\sim}{y})/p(\underset{\sim}{\beta})$ which fills the role of a pseudo-likelihood and represents the contribution of information about $\underset{\sim}{\beta}$ coming from the data. This factor is the posterior distribution of β when the prior is taken to be uniform. With this notation then for the present example

$$p(\underset{\sim}{\theta}|\underset{\sim}{y}) = \int_{-1}^{1} p(\underset{\sim}{\theta}|\beta,\underset{\sim}{y})p_u(\beta|\underset{\sim}{y})p(\beta)d\beta = \int_{-1}^{1} p_u(\underset{\sim}{\theta},\beta|\underset{\sim}{y})p(\beta)d\beta. \tag{14}$$

For Darwin's data the distributions $p_u(\theta,\beta|y)$ and a particular $p(\beta)$ are shown in Figure 3. Figure 4 shows

---

*Notice however, the distinction that must be drawn between criterion and inference robustness [6], [14]. For example, for these data the <u>significance level</u> of the <u>t criterion</u> is changed hardly at all (from 2.485% to 2.388%) if we suppose the parent distribution is rectangular rather than normal.

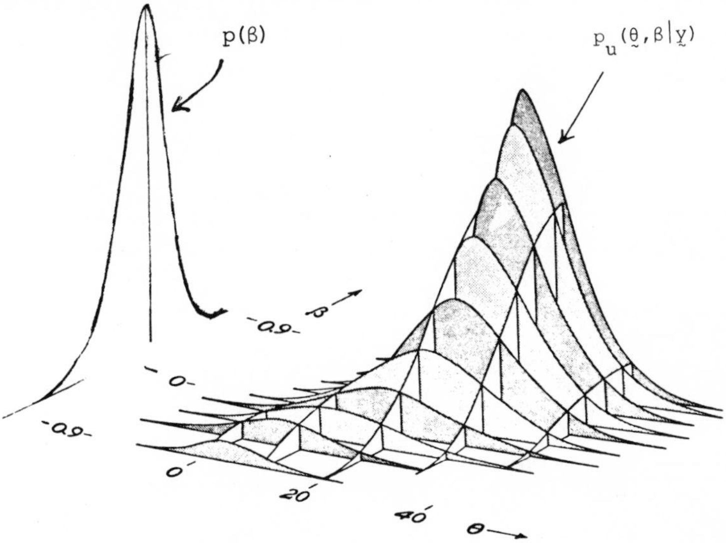

Figure 3. Joint posterior distribution $p_u(\theta, \beta | y)$ with a particular prior $p(\beta)$. Darwin's data.

$p(\beta | \underset{\sim}{y})$ for various choices of $p(\beta)$ while Figure 5 shows the corresponding distribution $p(\theta | \underset{\sim}{y})$.

a) Making $p(\beta)$ a delta function at $\beta = 0$ corresponds with the familiar absolute assumption of normality. It results in distribution (a) in Figure 5 which is the scaled t referred to earlier.

b) This choice for $p(\beta)$ is appropriate to a prior assumption that although not all distributions are normal; variations in kurtosis are such that the normal distribution takes a central role. For this particular example the resulting distribution (b) in Figure 5 is not very different from the t distribution.

c) Here, by making $p(\beta)$ uniform the modifier or pseudo-likelihood $p_u(\beta | \underset{\sim}{y})$ is explicitly produced which represents the information about kurtosis coming from the sample itself. For this extreme form of prior distribution, distribution (c) in Figure 5 is somewhat changed although not dramatically. The reason for this is that

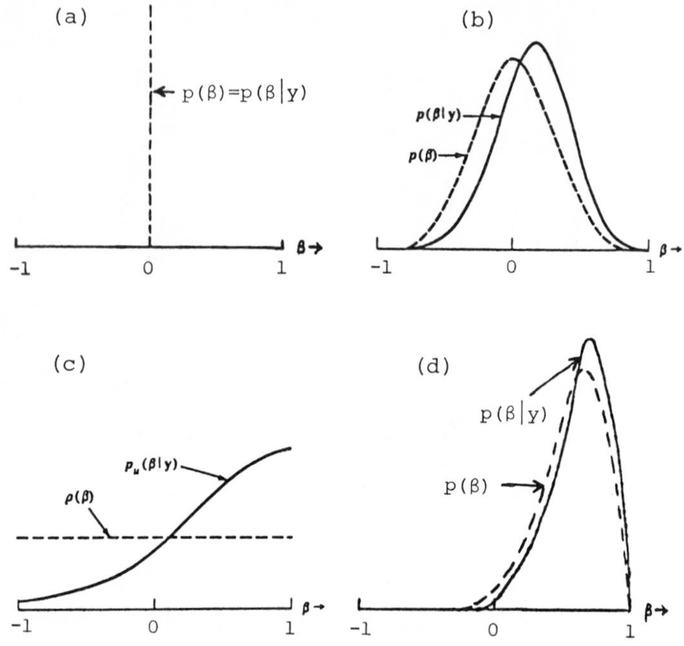

Figure 4. Posterior distributions $p(\beta|\underset{\sim}{y})$ for various choices of $p(\beta)$. Darwin's data.

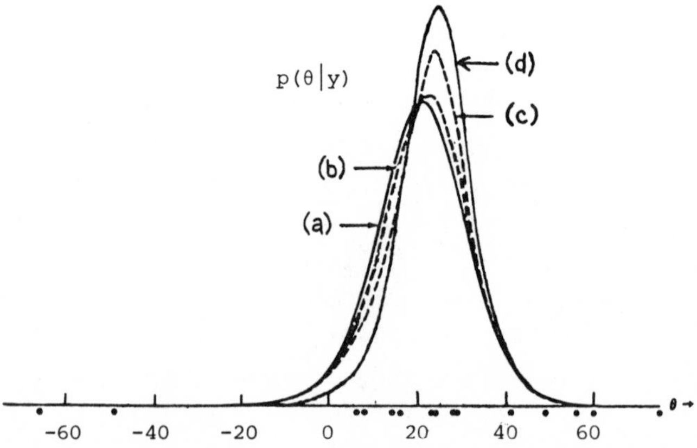

Figure 5. Posterior distributions $p(\theta|y)$ for various choices of $p(\beta)$. Darwin's data.

the widely discrepant distributions in Figure 2 for negative values of β are discounted by the information coming from the data.

d) This distribution is introduced to represent the kind of prior ideas which, following Tukey, many current robustifiers seem to have adopted. The resulting posterior distribution is shown in Figure 5(d).

This example brings to our attention the potential importance even for small samples of information coming from the data about the parameters $\tilde{\beta}$. In general if we compute the modifier $p_u(\tilde{\beta}|\tilde{y})$ from

$$p_u(\tilde{\beta}|\tilde{y}) = \frac{\int p(\tilde{\theta},\tilde{\beta}|\tilde{y})d\tilde{\theta}}{p(\tilde{\beta})} \tag{15}$$

then we can write

$$p(\tilde{\theta}|\tilde{y}) = \int p(\tilde{\theta}|\tilde{\beta},\tilde{y})p_u(\tilde{\beta}|\tilde{y})p(\tilde{\beta})d\tilde{\beta} \; . \tag{16}$$

Now even when the sensitivity factor is high that is when $p(\tilde{\theta}|\tilde{\beta},\tilde{y})$ changes rapidly as $\tilde{\beta}$ changes, this will lead to no uncertainty about $p(\tilde{\theta}|\tilde{y})$ if $p(\tilde{\beta}|\tilde{y})$ is sharp. This can be so either if $p(\beta)$ is sharp - there is an absolute assumption that we know β a priori - or if $p_u(\beta|\tilde{y})$ is sharp. In the common situation, the spread of $p_u(\beta|\tilde{y})$ will be proportional to $1/\sqrt{n}$ and for sufficiently large samples there will be a great deal of information from the sample about the relevant discrepancy parameters β. For small samples however this is not generally so. This amounts to saying that for sufficiently large samples it is always possible to check assumptions and in principle to robustify by incorporating sample information about discrepancies β into our statistical procedure. For small samples we are always much more at the mercy of the accuracy of prior information whether we incorporate it by using Bayes theorem or not. Notice however, how a Bayes analysis can make use of sample data which would have been neglected by a sampling theory analysis. Comparison of Figures 2 and 5(c) makes clear the profound effect that the sampling information about β for only $n = 15$ observations has on the inferential situation. This sample information represented by $p_u(\beta|\underset{\sim}{y})$ in Figure 4(c), although vague, is effective in discounting the possibility of platykurtic distributions which are the major cause of discrepancy in Figure 2. This effect accounts for the very moderate changes that occur in $p(\theta|\underset{\sim}{y})$ accompanying the drastic changes made in $p(\beta)$.

EXAMPLE 2: SERIAL CORRELATION AND REGRESSION ANALYSIS

Coen, Gomme and Kendall [9] gave 55 quarterly values of the Financial Times ordinary share index y_t, U.K. car production X_{1t} and Financial Times commodity index X_{2t}. They related y_t to the lagged values X_{1t-6} and X_{2t-7} by a regression equation

$$y_t = \theta_0 + \theta_1 t + \theta_2 X_{1t-6} + \theta_3 X_{2t-7} + \varepsilon_t \tag{17}$$

which they fitted by least squares. As mentioned earlier they obtained estimates of θ_2 and θ_3 which were very highly significantly different from zero and concluded that X_1 and X_2 could be used as "leading indicators" to predict future share prices. Box and Newbold [10] pointed out that if allowance is made for the serial correlation which exists in the error ε_t then the apparently significant effects vanish and much better forecasts are obtained by using today's price to forecast the future. This is a case where inferences about $\underset{\sim}{\theta}$ are very non-robust to possible serial correlation.

In a recent Wisconsin Ph.D. thesis [15] Lars Pallesen reassessed the situation with a Bayesian analysis, supposing that ε_t may follow a first order autoregressive model $\varepsilon_t - \beta\varepsilon_{t-1} = a_t$, where a_t is a white noise sequence as earlier defined.

The dramatic shifts that occur in the posterior distributions of θ_2 and θ_3 when it is not assumed a priori that $\beta = 0$ are shown in Figures 6 and 7. The situation is further illuminated by Figures 8 and 9 which show the joint distribution of θ_2 and β and of θ_3 and β, together with the marginal distribution $p_u(\beta|y)$ based on non-informative prior distributions.

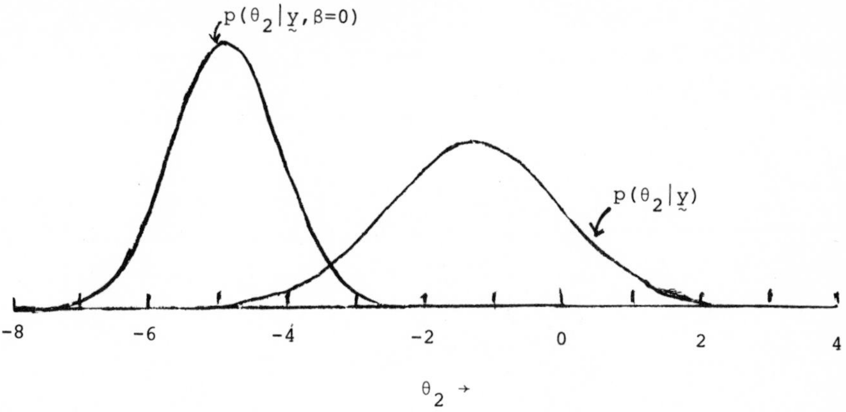

Figure 6. Effect of different assumptions on posterior distribution of θ_2. $p(\theta_2|\underset{\sim}{y})$ allows for possible autocorrelation of errors $p(\theta_2|y,\beta = 0)$ does not. θ_2 is regression coefficient of Share Index on car sales lagged six quarters.

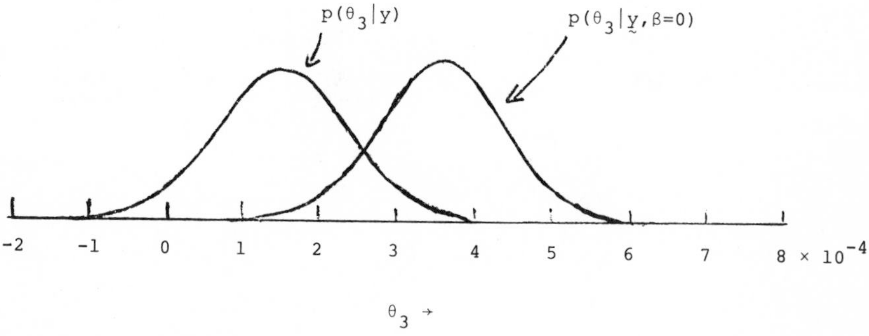

Figure 7. Effect of different assumptions on posterior distribution of θ_3. $p(\theta_3|\underset{\sim}{y})$ allows for possible autocorrelation of errors $p(\theta_3|y,\beta = 0)$ does not. θ_3 is regression coefficient of Share Index on Consumer Price Index lagged seven quarters.

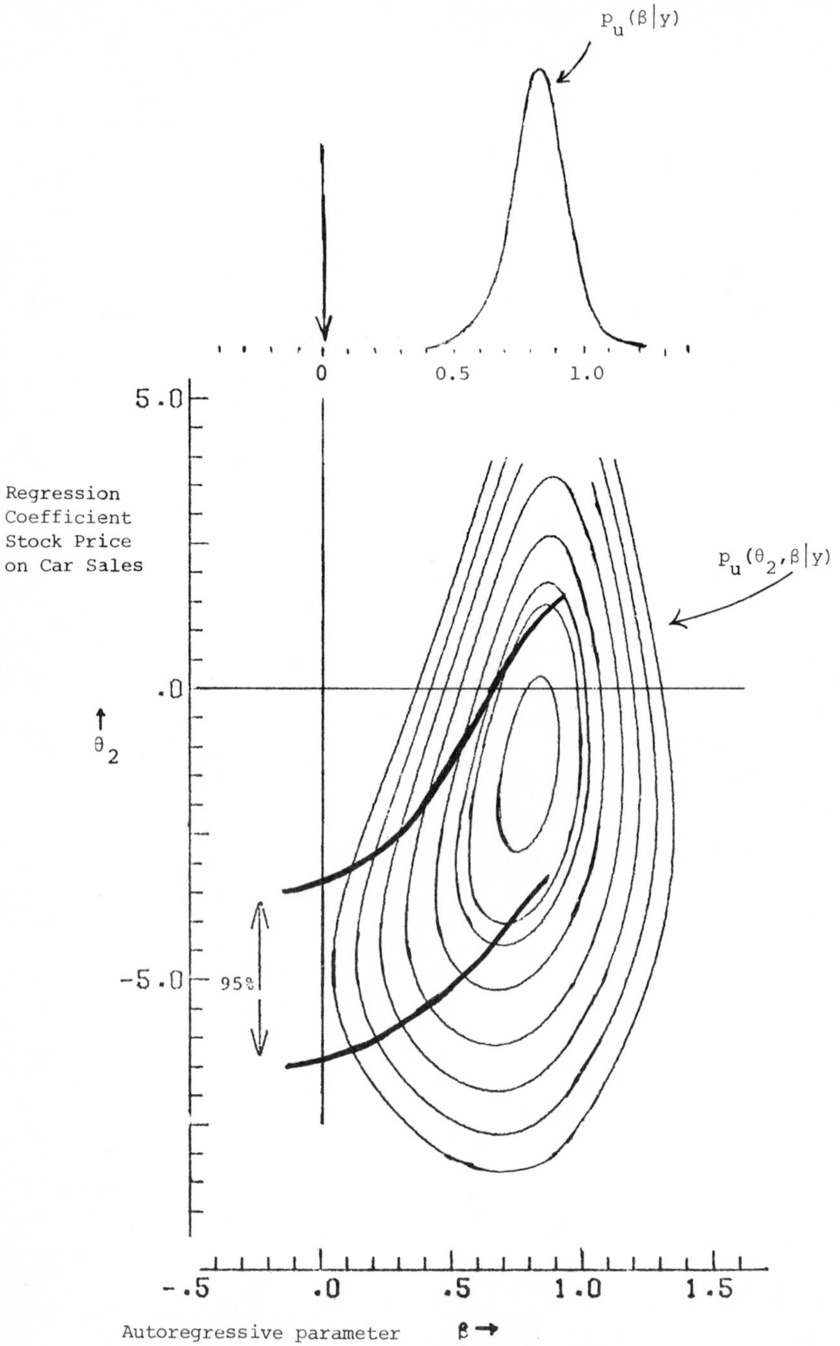

Figure 8. Joint posterior distribution of θ_2 and β and marginal posterior distribution of β. Note shift in 95% interval as β is changed.

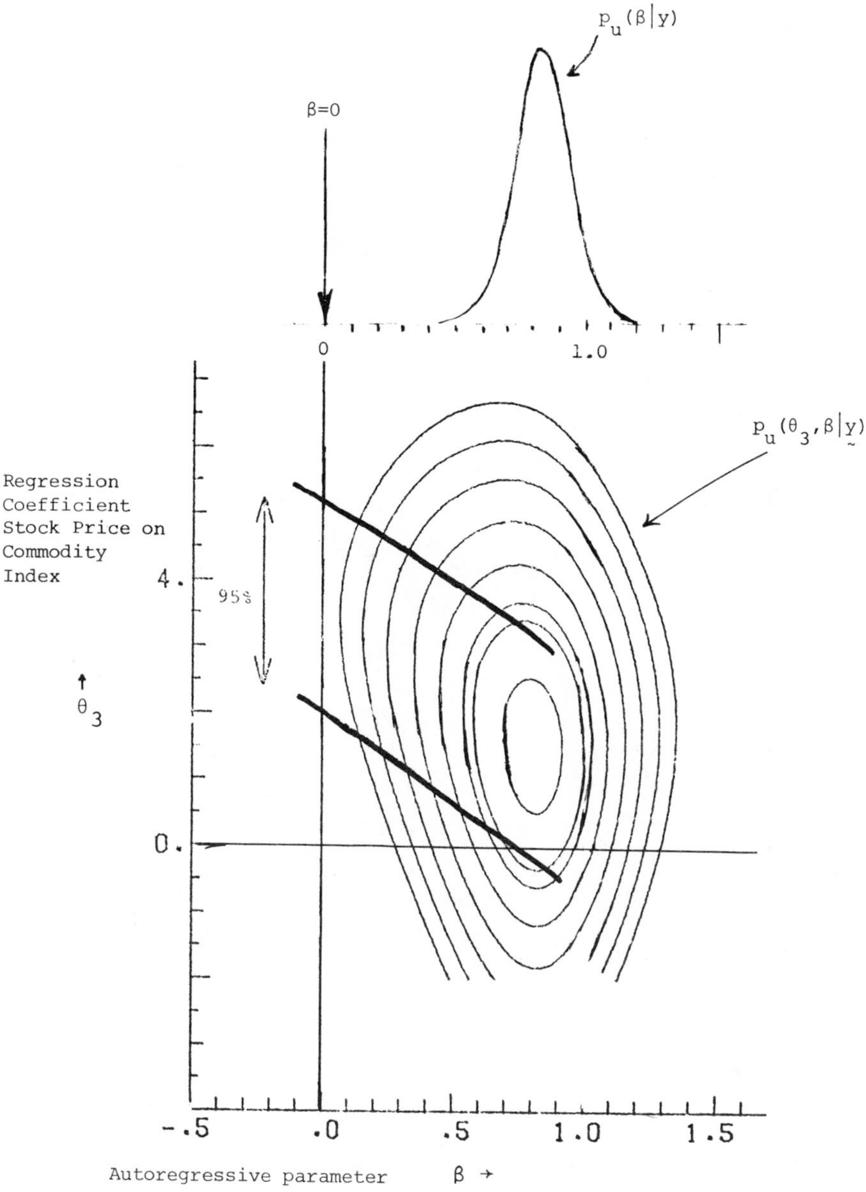

Figure 9. Joint posterior distribution of θ_3 and β and marginal posterior distribution of β. Note shift in 95% interval for θ_3 as β is changed.

EXAMPLE 3: OUTLIERS IN STANDARD STATISTICAL MODELS

Consider again the model form (1)

$$y_u = f(\underset{\sim}{\xi}_u, \underset{\sim}{\theta}) + \varepsilon_u \qquad u = 1,2,\ldots,n . \tag{18}$$

With standard assumptions about ε_u with the restriction that the expectation function is linear in $\underset{\sim}{\theta}$ this is the widely used Normal linear model. The remarkable thing about this model is that it is ever seriously entertained even when assumptions of independence and normality seem plausible. For it specifically states that the model form is appropriate for $u = 1,2,\ldots,n$ (that is, for every one of the experiments run). Now anyone who has any experience of reality knows that data are frequently affected by accidents of one kind or another resulting in "bad values". In particular it is expecting too much of any flesh and blood experimenter that he could conduct experiments unerringly according to a pre-arranged plan. Every now and again at some stage in the generation of data, a mistake will be made which is unrecognized at the time. Thus a much more realistic statement would be that model like (18) applied, not for $u = 1,2,\ldots,n$, but in a proportion $1 - \alpha$ of the time and that during the remaining proportion α of the time some discrepant, imprecisely known, model was appropriate. Such a model was proposed by Tukey in 1960 [16]. We call observations from the first model "good" and those from the second model "bad".

This type of model was later used in a Bayesian context by Box and Tiao [17]. They assumed that the discrepant model which generated the bad values was of the same form as the standard model except that the error standard deviation was k times as large. The results are rather insensitive to the choice of k. A Bayesian analysis was later carried out by Abraham and Box [18] with a somewhat different version of the model which assumes that the discrepant errors contain an unknown bias δ.

Either approach yields results which are broadly similar in that the posterior distribution of the parameters $\underset{\sim}{\theta}$ appears as a weighted sum of distributions.

$$p(\theta|y) = w_o p_o(\theta|y) + \sum_{i=1}^{n} w_i p_i(\theta|y) + \sum_{i=j}^{n} w_{ij} p_{ij}(\theta|y) + \ldots \tag{19}$$

The distribution $p_o(\underset{\sim}{\theta}|\underset{\sim}{y})$ in the first term on the right would be appropriate if all n observations were good, i.e. generated from the central model. The distribution $p_i(\underset{\sim}{\theta}|\underset{\sim}{y})$ in the first summation allows the possibility that $n - 1$ observations are good but the ith is bad. The next summation allows for two bad observations and so on. The weights w are posterior probabilities, w_o that no observation is bad, w_i that only the ith observation is bad, w_{ij} that the ith and jth observations are bad and so on.

Strictly the series includes all 2^n possibilities but in practice terms after the first or second summation usually become negligible.

Figure 10 shows an analysis for the Darwin data mentioned earlier. In this analysis it is supposed that the error distribution for good values is $N(0,\sigma^2)$ and that for bad values is $N(0,k^2\sigma^2)$. The analysis is made, as before, using a non-informative prior for θ and σ with $k = 3$ and $\alpha = .05$. This choice of α is equivalent to supposing that with 20 observations there is a 63.2% chance that one or more observations are bad. The probability of at least one outlier for other choices of n and α are given below

	α		
n	0.10	0.05	0.01
10	63.2	39.2	9.5
15	77.7	52.8	13.9
20	86.5	63.2	18.1
40	98.0	86.5	33.0

The results [17] are very insensitive to the choice of k but are less insensitive to the choice of α. However

1) it should be possible for the investigator to guess this value of α reasonably well.
2) the calculation can be carried out for different α values and the effect of different choices considered [18].

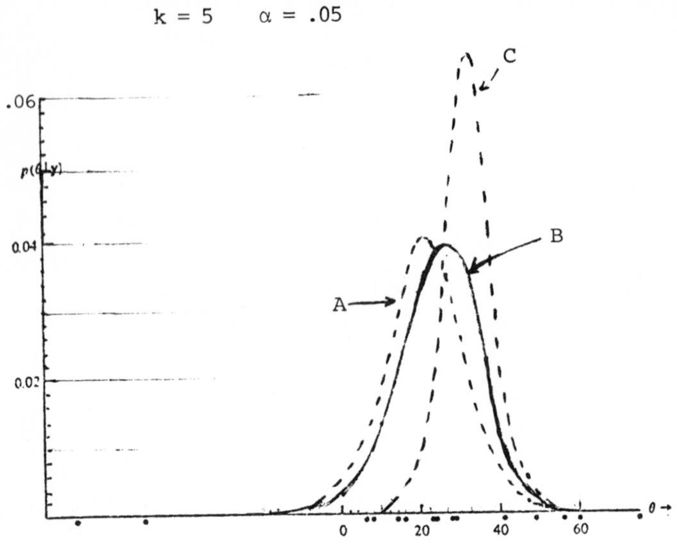

Figure 10. A. Assuming no outliers.
B. Allowing the possibility of outliers.
C. Assuming y_1 and y_2 <u>are</u> outliers.

Inspection of the weights w can also be informative in indicating possible outliers. For example [19] the time series shown in Figure 11 consists of 70 observations generated from the model:

$$y_t = \phi y_{t-1} + \delta_t + a_t$$

where

$$\delta_t = \begin{cases} 5 & \text{if } t = 50 \\ 0 & \text{otherwise} \end{cases} \tag{20}$$

$\phi = .5$ and $\{a_t\}$ a set of independent normally distributed random variables with variance $\sigma_a^2 = 1$. The plot in Figure 12 of the weights w_i indicates the probability of each being bad and clearly points to discrepancy of the 50th observation.

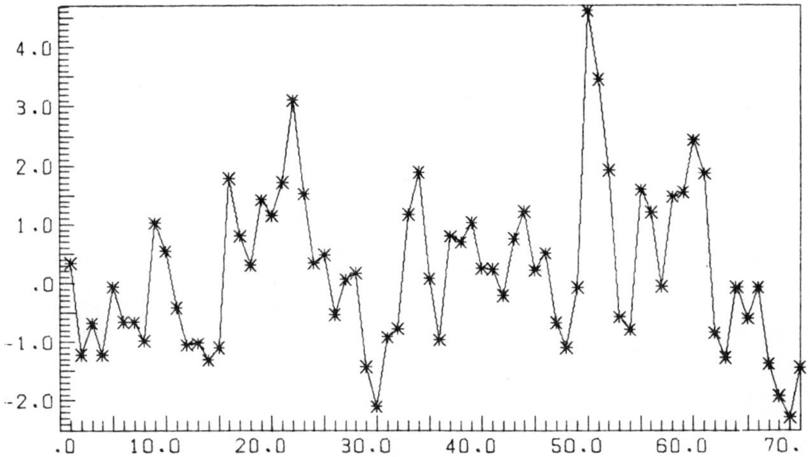

Figure 11. A time series generated from a first order autoregressive model with an outlier innovation at $t = 50$.

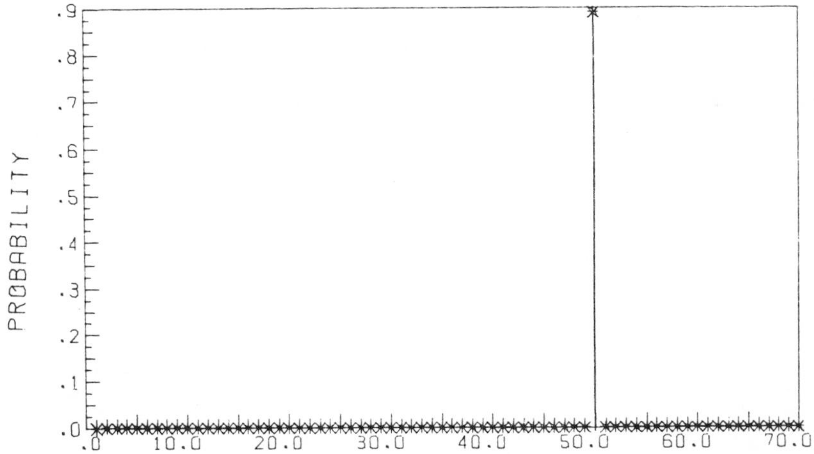

Figure 12. Posterior probabilities of bad values given that there is one bad value.

EXAMPLE 4: TRANSFORMATION OF THE DEPENDENT VARIABLE

Parsimony favors any device that expands model applicability with small expenditure of additional parameters. As was emphasized by Fisher, in suitable circumstances, parametric transformation can provide such a device. For example a suitable power transformation $Y = y^{\lambda}$ can have profound effect when y_{max}/y_{min} is not small.

In this application then the discrepancy parameter β measures the need for transformation. In particular for the power transformations no transformation corresponds to $\beta = \lambda = 1$.

The Bayes approach to parametric transformations was explored by Box and Cox [11]. One example they considered concerned a 3×4 factorial design with 4 animals per cell in which a total of $n = 48$ animals were exposed to three different poisons and four different treatments. The response was survival time. Since for this data $y_{max}/y_{min} = 12.4/1.8 \simeq 7$ we know a priori that the effect of needed transformation could be profound and it would be sensible to make provision for it in the first tentative model.

For this particular set of data, where there is a blatant need for transformation, an initial analysis with no transformation followed by iterative fix up would be effective also. Diagnostic checks involving residual plots of the kind suggested by Anscombe and Tukey [20], [21] certainly indicate [22] the dependence of cell variance on cell mean and less clearly non-additivity. Whatever route we take we are led to consider a transformation y^{λ} where λ approaches -1. As will be seen from the analysis of variance below this transformation not only eliminates any suggestion of an interaction between poisons and treatments but also greatly increases precision. This example seems to further illustrate how Bayesian robustification of the model illuminates the relation of the data to a spectrum of models. Using noninformative prior distributions Figure 13 shows posterior distributions for λ with different constraints applied to the basic normal, independent, model

$$y_{rci} = \mu_{rc} + \varepsilon_{rci} \tag{21}$$

where the subscripts r, c, i apply to rows, columns and replicates.

Analyses of Variance of the Biological Data

	Mean squares × 1000			
	Degrees of freedom	Untransformed	Degrees of freedom	Reciprocal transformation (z form)
Poisons	2	516.3	2	568.7
Treatments	3	307.1	3	221.9
P × T	6	41.7	6	8.5
Within groups	36	22.2	36(35)	7.8(8.0)

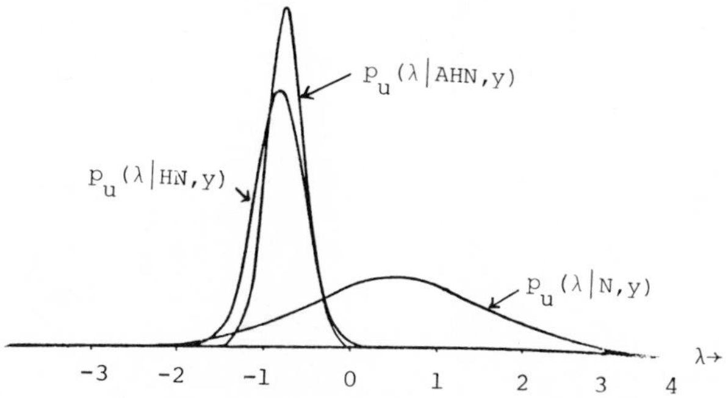

Figure 13. Posterior distributions for λ under various constraints N-Normality, Homogeneity of variance, A-Additivity.

The nature of the various distributions is indicated in the following table in which N, H, and A refer respectively to Normality, Homogeneity of Variance and Additivity and ρ_r and γ_c are row and column effects

Distribution	Constraint
$p_u(\lambda\|N,y)$	$V(\varepsilon_{rci}) = \sigma^2_{rc}$
$p_u(\lambda\|HN,y)$	$\sigma^2_{rc} = \sigma^2$
$p_u(\lambda\|AHN,y)$	$\mu_{rc} = \mu + \rho_r + \gamma_c$ and $\sigma^2_{rc} = \sigma^2$.

The disperse nature of the distribution $p_u(\lambda|N,y)$ is to be expected since a sample of size $n = 48$ cannot tell us much about normality. The greater concentration of $p_u(\lambda|HN,y)$ arises because there is considerable variance heterogeneity in the original metric which is corrected by strong transformations in the neighborhood of the reciprocal. Finally $p(\lambda|AHN)$ is even more concentrated because transformations of this type also remove possible non-additivity. The analysis suggests among other things that for this data the choices of transformations yielding approximate additivity, homogeneity of variance and normality are not in conflict. From Figure 14 (taken from [22]) we see that for this example appropriate adjustment of the discrepancy parameter $\beta = \lambda$ affects not only the location of the poison main effects, it also has a profound effect on their precision*. Indeed the effect of including λ in estimating main effects is equivalent to increasing the sample size by a factor of almost three.

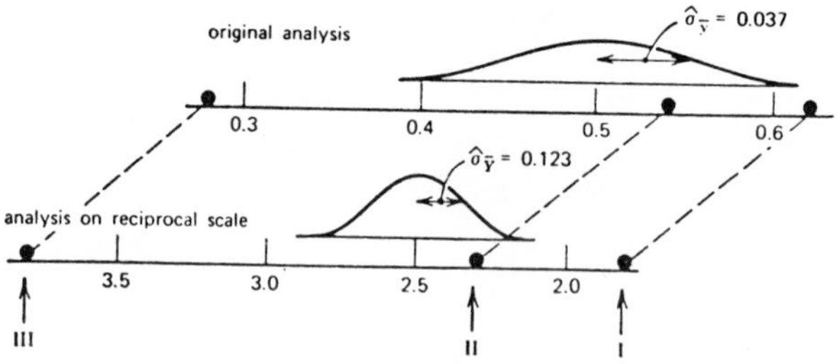

Figure 14. Posterior distributions for individual means (poisson main effects) on original and reciprocal scale. Note greatly increased precision induced by appropriate transformation.

PSYCHING OUT THE ROBUSTIFIERS

To apply Bayesian analysis we must choose a $p(\underset{\sim}{\beta})$ which roughly describes the world we believe applies in the problem context.

There are a number of ways we can be assisted in this choice.

(1) We can look at extensive sets of data and build up suitable distributions $p(\underset{\sim}{\beta})$ from experience.

---

*Similar results are obtained for the treatment effects and as noted before the transformation eliminates the need for interaction parameters.

(ii) We can consult experts who have handled a lot of data of the kind being considered.
(iii) We can consider the nature of the robust estimates proposed and what they reveal about the proposer's prior beliefs.

Consider, for example, the heavy tailed error problem. Gina Chen in a recent Ph.D. thesis [23] has found prior distributions $p(\underset{\sim}{\beta})$ yielding posterior means which approximate robust estimates of location already proposed on other grounds. In one part of her study she considers a model in which data come from an exponential power distribution. $p(y|\theta,\sigma,\beta)$ of the form of (11) with probability $1 - \alpha$, and, with probability α to have come from a similar distribution but with a standard deviation k times as large. Thus

$$p(y|\theta,\sigma,\alpha,\beta) = (1 - \alpha)p(y|\theta,\sigma,\beta) + \alpha\cdot p(y|\theta,k\sigma,\beta) \ . \tag{22}$$

It turns out in fact that priors which put all the mass at individual points in the $p(\alpha,\beta)$ plane can very closely approximate suggested M estimators as well as trimmed means, Winsorized means, and other L estimators.

Three objectives of her study were
(i) To make it possible to examine more closely and hence to criticize the assumptions about the real world which would lead to the various robust estimates.
(ii) To compare these revealed assumptions with the properties of actual data.
(iii) To allow conclusions obtained from simple problems to be applied more generally. Once we agree on what $p(\underset{\sim}{\beta})$ should be for a location parameter then the same $p(\underset{\sim}{\beta})$ can be used for more complicated problems occurring in the same scientific context. Direct application of Bayes theorem can then, for example, indicate the appropriate analysis for all linear and nonlinear models formerly analyzed by least squares.

SUMMARY AND CONCLUSIONS

A major activity of statisticians should be to help the scientist in his iterative search for useful but necessarily inexact parsimonious models. While inexact models may mislead, attempting to allow for every contingency a priori is impractical. Thus models must be built by an iterative feedback process in which an initial parsimonious model may be modified when diagnostic checks applied to residuals indicate the need.

When discrepancies may occur which are unlikely to be detected by diagnostic checks, this feedback process could fail and therefore procedures must be robustified with respect to these particular kinds of discrepancies. This writer believes that this may best be done by suitably modifying the model rather than by modifying the method of inference.

In particular a Bayes approach offers many advantages. Suppose the scientist wishes to protect inferences about <u>primary parameters</u> $\underset{\sim}{\theta}$ from effects of <u>discrepancy parameters</u> $\underset{\sim}{\beta}$. Bayes analysis automatically brings into the open a number of important elements.

(i) The prior distribution $p(\underset{\sim}{\beta})$ reveals the nature of the supposed universe of discrepancies from which the procedure is being protected.

(ii) The distribution $p_u(\underset{\sim}{\beta}|\underset{\sim}{y}) = p(\underset{\sim}{\beta}|\underset{\sim}{y})/p(\underset{\sim}{\beta})$ represents information about $\underset{\sim}{\beta}$ coming from the data itself. This distribution may be inspected for concordance with $p(\underset{\sim}{\beta})$.

(iii) The conditional posterior distribution $p(\underset{\sim}{\theta}|\underset{\sim}{\beta},\underset{\sim}{y})$ shows the sensitivity of inferences to choice of $\underset{\sim}{\beta}$.

(iv) From the marginal posterior distribution $p(\underset{\sim}{\theta}|\underset{\sim}{y})$ appropriate inferences which are robust with respect to $\underset{\sim}{\beta}$ may be made.

(v) Implications of inspired empiricism can lead to useful models. For example, we can ask "What kind of $p(\underset{\sim}{\beta})$ will make some empirical robust measure of location a Bayesian estimator?" This $p(\underset{\sim}{\beta})$ may then be examined, criticized and perhaps compared with distributions of $\underset{\sim}{\beta}$ encountered in the real world.

(vi) Once $p(\underset{\sim}{\beta})$ is agreed on then that same $p(\underset{\sim}{\beta})$ can be applied to other problems. For example, we do not need to give special consideration to robust regression, robust analysis of variance, robust non-linear estimation. We simply carry through the Bayesian analysis with the agreed $p(\underset{\sim}{\beta})$.

(vii) In the past the available capacity and speed of computers might have limited this approach but this is no longer true. It will be necessary however, to make a major effort to produce suitable programs which can readily perform analyses and display results of the kind exemplified in this paper.

APPENDIX 1

Suppose that, in model (1), n observations are available and standard assumptions of independence and homoscedasticity are made about the errors $\{\varepsilon_u\}$. Suppose finally that the object is to estimate $E(y)$ over a region in the space of $\underset{\sim}{\xi}$ "covered" by the experiments $\{\underset{\sim}{\xi}_u\}$. Then the number of parameters p employed in the expectation function is a natural measure of prodigality and its reciprocal $1/p$ of parsimony.

Now denote by $\hat{y}_u^{(p)} = f^{(p)}(\underset{\sim}{\xi}_u, \hat{\underset{\sim}{\theta}}_p)$ a fitted value with estimates $\hat{\underset{\sim}{\theta}}_p$ obtained by least squares and by

$$\bar{V}^{(p)} = \sum_u V\{\hat{y}_u^{(p)}\}/n \qquad (A.1)$$

the average variance of the n fitted values.

It is well known that (exactly if the expectation function is linear in $\underset{\sim}{\theta}$, and in favorable circumstances, approximately otherwise) no matter what experimental design $\{\underset{\sim}{\xi}_u\}$ is used

$$\bar{V}^{(p)} = p\sigma^2/n \ . \qquad (A.2)$$

Now if the $\{\eta_u\}$ can be regarded as a sampling of the function over the region of interest, then $\bar{V}^{(p)}$ provides a measure of average variance of estimate of the function over the experimental region.

Equation (A.2) says that this average variance of estimate of the function is proportional to the prodigality p. Alternatively it is reasonable to regard $I^{(p)} = \{\bar{V}^{(p)}\}^{-1}$ as a measure of information supplied by the experiment about the function and

$$I^{(p)} = n/p\sigma^2 \ . \qquad (A.3)$$

Thus this measure of information is proportional to the parsimony $1/p$. For example, if the expectation function needed as many parameters as there were observations so that $p = n$ then $\hat{y}_t = y_t$ and

$$\bar{V}^{(n)} = V(y_t) = \sigma^2 \qquad I^{(n)} = 1/\sigma^2 . \tag{A.4}$$

In this case the model does not summarize information and does not help in reducing the variance of estimate of the function.

At the other extreme if the model needed to contain only a single parameter, for example,

$$y_t = \theta + \varepsilon_t \tag{A.5}$$

then $\hat{\theta} = \hat{y}_t = \bar{y}$ and

$$\bar{V}^{(1)} = V(\bar{y}) = \sigma^2/n \qquad I^{(1)} = n/\sigma^2 . \tag{A.6}$$

In this case the use of the model results in considerable summarizing of information and reduces the variance of estimate of the function n times or equivalently increases the information measure n-fold.

Considerations of this sort weigh heavily against unnecessarily complicated models.

As an example of unnecessary complication consider an experimenter who wished to model the deviation $\tilde{y}_t$ from its mean of the output from a stirred mixing tank in terms of the deviation $\tilde{\xi}_t$ from its mean of input feed concentration. If data were available to equal intervals of time, he might use a model

$$\tilde{y}_t = \{\theta_o\tilde{\xi}_t + \theta_1\tilde{\xi}_{t-1} + \theta_2\tilde{\xi}_{t-2} + \cdots + \theta_k\tilde{\xi}_{t-k}\} + \varepsilon_t \tag{A.7}$$

in which k was taken sufficiently large so that deviations in input ξ_{t-k-q} for $q > 0$ were assumed to have negligible effect on the output at time t. This model contains $k + 1$ parameters $\underset{\sim}{\theta}$ which need to be estimated. Alternatively if he knew something about the theory of mixing he might instead tentatively entertain the model

$$\tilde{y}_t = \theta_o\{\tilde{\xi}_t + \theta_1\tilde{\xi}_{t-1} + \theta_1^2\tilde{\xi}_{t-2} + \cdots\} + \varepsilon_t \tag{A.8}$$

or equivalently

$$\tilde{y}_t = \theta_1 \tilde{y}_{t-1} + \theta_o \tilde{\xi}_t + \varepsilon_t - \theta_1 \varepsilon_{t-1} \qquad \text{(A.9)}$$

which contains only two parameters $\underset{\sim}{\theta}$.

Thus if the simpler model provided a fair approximation, it could result in greatly increased precision as well as understanding.

APPENDIX 2

The practical importance of worrying about the right things is illustrated, for example, by the entries in the following table taken from [7], [22]. This shows the results of a sampling experiment designed to compare the robustness to non-normality and to serial correlation of the significance level of the t test and the non-parametric Mann-Whitney test. One thousand pairs of samples of ten of independent random variables u_t were drawn from a rectangular distribution, a normal distribution and a highly skewed distribution (a χ^2 with 4 degrees of freedom) all adjusted to have mean zero. In the first row of the table the errors $\varepsilon_t = u_t$ were independently distributed, in the second and third rows a moving average model $\varepsilon_t = u_t - \theta u_{t-1}$ was used to generate errors with serial correlation -0.4 and $+0.4$ respectively. The numbers on the right show the corresponding results when the pairs of samples were randomized.

In this example the performance characteristic studied is the numbers of samples showing significance at the 5% level when the null hypotheses of equality of means was in fact true. Under ideal assumptions the number observed would, of course, vary about the expected value of 50 with a sampling standard deviation of about 7. It is not intended to suggest by this example that the performance of significance tests when the null hypothesis is true is the most important thing to be concerned about. But rightly or wrongly, designers of non-parametric tests have been concerned about it, and demonstrations of this kind suggest that their labors are to some extent misdirected. In this example it is evident that it is the physical act of randomization and much less so the choice of criterion that protects the significance level.

TABLE A.1

Tests of Two Samples of Ten Observations Having the Same Mean. Frequency in 1,000 Trials of Significance at the 5 Percent Level Using the t-Test (t) and the Mann-Whitney Test

			With No Randomization			After Randomization		
	ρ_1	Test	Rectangular	Normal	Chi-square*	Rectangular	Normal	Chi-square*
Independent Observations	0.0	t	56	54	47	60	43	49
		MW	43	45	43	58	41	44
Autocorrelated Observations	-0.4	t	5	3	1	48	55	63
		MW	5	1	2	43	49	56
	0.4	t	125	105	114	59	58	54
		MW	110	96	101	46	53	43

*This parent chi-square distribution has four degrees of freedom and is thus highly skewed.

REFERENCES

1. Box, G.E.P. and Jenkins, G.M. (1970). Time Series Analysis: Forecasting and Control, Holden-Day.
2. Box, G.E.P., Hillmer, S.C. and Tiao, G.C. (1976). Analysis and Modelling of Seasonal Time Series. Technical Report No. 465, Department of Statistics, University of Wisconsin, Madison.
3. Watson, James D. (1968). The Double Helix. New York: Atheneum. A personal account of the discovery of the structure of DNA.
4. Box, J.F. (1978). R.A. Fisher, The Life of a Scientist. Wiley, Chapter 7.
5. Box, G.E.P. and Youle, P.V. (1955). "The Exploration and Exploitation of Response Surfaces: An Example of the Link Between the Fitted Surface and the Basic Mechanism of the System", Biometrics 11, 287.
6. Box, G.E.P. and Tiao, G.C. (1973). Bayesian Inference in Statistical Analysis. Addison-Wesley, p. 305, Chapter 5.
7. Box, G.E.P. (1976). "Science and Statistics", JASA 71, 791.
8. Tukey, J.W. (1949). "One Degree of Freedom for Non-additivity", Biometrics 5, 232.
9. Coen, P.G., Gomme, E.D. and Kendall, M.G. (1969). "Lagged Relationships in Economic Forecasting", JRSS, Series A 132, 133.
10. Box, G.E.P. and Newbold, Paul (1971). "Some Comments on a Paper of Coen, Gomme, and Kendall, JRSS, Series A 134, 229.
11. Box, G.E.P. and Cox, D.R. (1964). "An Analysis of Transformations", JRSS, Series B 26, 211.
12. Tiao, G.C. and Box, G.E.P. (1973). "Some Comments on "Bayes" Estimators", The American Statistician 27, No. 1, 12.
13. Box, G.E.P. and Tiao, G.C. (1962). "A Further Look at Robustness via Bayes's Theorem", Biometrika 49, 419.
14. Box, G.E.P. and Tiao, G.C. (1964b). "A Note on Criterion Robustness and Inference Robustness", Biometrika 51, 169.

15. Pallesen, L.C. (1977). "Studies in the Analysis of Serially Dependent Data", Ph.D. thesis, the University of Wisconsin, Madison.

16. Tukey, J.W. (1960). "A survey of sampling from contaminated distributions" in Contributions to Probability and Statistics, Stanford University Press, 448-485.

17. Box, G.E.P. and Tiao, G.C. (1968). "A Bayesian Approach to Some Outlier Problems", Biometrika 55, 119.

18. Abraham, B. and Box, G.E.P. (1978). "Linear Models and Spurious Observations", Appl. Statist. 27, 131-138.

19. Abraham, B. and Box, G.E.P. (1979). "Outliers in Time Series", (to appear) Biometrika, see also Technical Report #440, Department of Statistics, University of Wisconsin, Madison.

20. Anscombe, F.J. (1961). "Examination of Residuals", Proceedings of 4th Berkeley Symposium, Math. Statist. Proc. 1, 1.

21. Anscombe, F.J. and Tukey, W.J. (1963). "The Examination and Analysis of Residuals", Technometrics 5, 141.

22. Box, G.E.P., Hunter, W.G. and Hunter, J.S. (1978). "Statistics for Experimenters", John Wiley.

23. Chen, Gina (1979). "Studies in Robust Estimation", Ph.D. thesis, the University of Wisconsin, Madison.

Sponsored by the United States Army under Contract No. DAAG29-75-C-0024.

Department of Statistics and
Mathematics Research Center
University of Wisconsin-Madison
Madison, WI 53706

A Density–Quantile Function Perspective on Robust Estimation

Emmanuel Parzen

1. INTRODUCTION.

There seems to be a growing consensus among statisticians that the discipline of statistics will be very different in the next decade from what it was in the last decade. There is less agreement on what the new shape will be. This paper attempts to provide a perspective on robust estimation by regarding it as one of the components of a unified theory of statistical science (including statistical data analysis), in which various approaches are applied simultaneously: parametric statistical inference, goodness of fit procedures, robust statistical inference, exploratory data analysis, and non-parametric statistical inference. It is proposed that a key to accomplishing this unification is to think in terms of quantile functions and density-quantile functions.

Parametric statistical inference may be said to be concerned with statistical inference of parameters of a model from data assumed to satisfy the model. Huber (1977), p. 1, writes: "The traditional approach to theoretical statistics was and is to optimize at an idealized parametric model."

Goodness of fit procedures are concerned with testing whether a specified parametric model adequately fits the observed sample probabilities.

ISBN 0-12-438150-2

Robust statistical inference may be said to be concerned with statistical inference of parameters of a model from data assumed to satisfy the model only approximately; in other words, statistical estimators are robust if they are insensitive against small deviations from ideal assumptions. Huber (1977), p. 3, writes: the robust approach to theoretical statistics assumes "an idealized parametric model, but in addition one would like to make sure that methods work well not only at the model itself, but also in a neighborhood of it."

Exploratory data analysis may be said to be concerned with statistical inference from data which is non-ideal in the sense that it is not assumed to obey a specified model. Often one seeks re-expressions (transformations) of the data that will make it ideal (obey a model).

It is proposed in this paper that exploratory data analysis (especially by the approach of non-parametric statistical data modeling) helps pose the well-posed statistical questions to which classical parametric statistics provides answers, and the vaguely-well-posed statistical questions to which robust statistics provides answers.

Point I of this paper consists of proposals for robust estimation of location parameters; this paper proposes that the means must be justified by the ends (by "ends" one means the tail character of the distribution of the data). Two basic models for robust estimation of location ("means") are:

(1) When the data distribution is diagnosed to be a symmetric possibly long tailed distributions, one uses robust estimators obtained as iteratively reweighted estimators with weight function $w(x) = (1 + \frac{1}{m}x^2)^{-1}$ for suitable choices of m (section 4).

(2) For contaminated normal data, one uses robust estimators obtained as maximum likelihood estimators omitting extreme order statistics, where the percentage of values omitted is determined from the goodness of fit of the corresponding smooth quantile functions (section 5).

Other methods of robust location and scale parameter estimation briefly indicated in section 6 are:

(3) Quantile box-plot diagnostics which indicate that mid-summaries and mid-scales are equal enough to provide naive estimators of location and scale.

(4) Adaptive L-estimation of location and scale parameters using autoregressive estimators of density-quantile functions given by Parzen (1978).

2. WHAT SHOULD BE TAUGHT IN AN INTRODUCTION TO STATISTICS?

Can statistics be made a subject that is regarded as relevant by the creative scientist, and is appealing as a career to the mathematically talented? An important step in achieving these desirable (and I believe attainable) goals is to alter the perception of the sample mean and the sample variance in elementary statistical instruction. Introductory Statistics is regarded by almost all college students (even by mathematically talented students) as a very dull subject. Perhaps one reason is that students enter the course knowing about a mean and a variance and leave the course knowing only about a mean and a variance. That statistics is in fact a live and vibrant discipline can be communicated to the student by emphasizing the new ways of thinking about estimators of mean and variance (and more generally of location and scale parameters) that require robust statistical procedures. The discipline of statistics needs to be made more "glamorous"; one way to accomplish this is to incorporate intellectually sound and demonstratively useful concepts of statistical data analysis and robust statistical inference in introductory statistical instruction.

The perspective which this paper proposes for interpreting robust statistical inference is equivalent to a proposal for a definition of statistics:

> "Statistics is arithmetic done by the method of Lebesgue integration."

I realize this definition sounds unbelievable and may never sell to the introductory students. But at least statisticians should understand to what extent it is true (perhaps it provides a basis for a new sect of statisticians).

Can we all agree that a basic problem of statistics is an arithmetical one: find the average $\overline{X}$ of a set of numbers $X_1, \ldots, X_n$? Even grade school students (in the U. S. A.) nowadays know the answer:

$$\overline{X} = \frac{1}{n}(X_1 + X_2 + \ldots + X_n) \quad .$$

In words: list the numbers, add them up, and divide by n . What should be realized is that the foregoing algorithm is the method of Riemann integration.

The method of Lebesgue integration finds $\overline{X}$ by first finding the distribution function $\tilde{F}(x)$ of the data, defined by $\tilde{F}(x)$ = fraction of $X_1, \ldots, X_n \leq x$, $-\infty < x < \infty$. Then $\overline{X}$ is found as the mean of this distribution function, defined by the integral

$$\overline{X} = \int_{-\infty}^{\infty} x d\tilde{F}(x) \quad .$$

To use an analogy: to count a sack of coins, first arrange the coins in piles according to their value (pennies, nickels, dimes, quarters, and half-dollars), then count the number of coins in each pile, determine the value of each pile, and finally obtain $\overline{X}$ as the sum of the values of the piles, divided by n . The role of statistics is to find more accurate estimators of the true mean by fitting a smooth distribution function $\hat{F}(x)$ to $\tilde{F}(x)$.

Still more insight (and fidelity to the truth) is obtained by displaying the sample quantile function $\tilde{Q}(u) = \tilde{F}^{-1}(u)$, defined in section 3, and ultimately by determining smooth quantile functions $\hat{Q}(u)$ which fit $\tilde{Q}(u)$. It is shown in section 4 that a "sample average" or an estimator of the "true average" should be computed by

$$\hat{\mu} = \int_0^1 W_\mu(u)\, \tilde{Q}(u)\, du \div \int_0^1 W_\mu(u)\, du \quad ,$$

for a suitable weight function $W_\mu(u)$, $0 \leq u \leq 1$.

In words, what the introductory student should learn is that the "average" of a sample is a weighted average of the numbers in the sample arranged in increasing order, with the weight of a number depending on its rank.

The achievement of robust estimation theory seems to me to have been to show that: (1) it is possible to choose the weights adaptively from the data so that they perform well under a variety of true distribution functions for the data, and in the presence of observations from contaminating distributions; (2) it is possible to extend robust procedures from problems of estimation of location and scale parameters to problems of estimation of linear regression coefficients; and (3) it is possible to preserve parameters as devices of statistical data analysis to summarize data (and relations between variables) because one can find estimators of such parameters with desirable properties for a variety of symmetric long-tailed or contaminated distributions of the data (or, for regression models, the errors).

Point II of this paper consists of the following proposals: (1) graphical goodness-of-fit procedures should be applied to data to check if they are adequately fitted by the qualitatively defined models which are implicitly assumed by robust estimation procedures; and (2) such procedures can be provided by Quantile-Box Plots and non-parametric estimators of the density-quantile function [Parzen (1978)].

3. QUANTILE FUNCTION.

The aim of this section is to introduce the quantile function and illustrate how it can be used to provide non-parametric measures of location (such as the median) and scale (such as the quartile spread). Quantile-Box Plots (defined in section 7) are an example of how to use quantile functions to detect and describe deviations from ideal statistical models for data.

The quantile function $Q(u)$, $0 \le u \le 1$ of a random variable X is the inverse of its distribution $F(x) = P[X \le x]$. The precise definition of Q is: $Q(u) = F^{-1}(u) = \inf\{x : F(x) \ge u\}$.

Given a sample $X_1, \ldots, X_n$, we denote the sample distribution function by $\tilde{F}(x)$, $-\infty < x < \infty$; it is defined by $\tilde{F}(x)$ = fraction of $X_1, \ldots, X_n \le x$.

The Sample Quantile Function $\tilde{Q}(u) = \tilde{F}^{-1}(u) = \inf\{x : \tilde{F}(x) \ge u\}$ can be computed explicitly in terms of the order statistics $X_{(1)} < X_{(2)} < \ldots < X_{(n)}$ (which are the values in the sample arranged in increasing order):

$$\tilde{Q}(u) = X_{(j)} \quad , \quad \frac{j-1}{n} < u \le \frac{j}{n} \quad . \tag{1}$$

The foregoing definition of $\tilde{Q}(u)$ is a piecewise constant function. It is more convenient to define $\tilde{Q}(u)$ as a piecewise linear function. Divide the unit interval into $2n$ subintervals. For $u = (2j - 1)/2n$ define

$$\tilde{Q}\left(\frac{2j-1}{2n}\right) = X_{(j)} \quad , \quad j = 1, 2, \ldots, n \quad . \tag{2}$$

For u in $\frac{2j-1}{2n} \le u \le \frac{2j+1}{2n}$, $j = 1, 2, \ldots, n-1$, define $\tilde{Q}(u)$ by linear interpolation; thus for u in this interval

$$\tilde{Q}(u) = n\left(u - \frac{2j-1}{2n}\right) X_{(j+1)} + n\left(\frac{2j+1}{2n} - u\right) X_{(j)} \quad . \tag{3}$$

In particular $\tilde{Q}\left(\frac{j}{2n}\right) = \frac{1}{2} X_{(j+1)} + \frac{1}{2} X_{(j)}$.

The population median is $Q(0.5)$. The sample median is $\tilde{Q}(0.5)$. Our definition of $\tilde{Q}(u)$ has the merit that $\tilde{Q}(0.5)$ is the usual definition of the sample median:

$$\begin{aligned} \tilde{Q}(0.5) &= X_{(m+1)} \quad \text{if} \quad n = 2m + 1 \quad \text{is odd,} \\ &= \frac{1}{2}(X_{(m)} + X_{(m+1)}) \quad \text{if} \quad n = 2m \quad \text{is even.} \end{aligned} \tag{4}$$

The asymptotic distribution of $\tilde{Q}(u)$ satisfies: $\sqrt{n}\, fQ(u)\{\tilde{Q}(u) - Q(u)\}$ is asymptotically normal, with mean 0 and variance $u(1 - u)$, where $fQ(u)$ denotes the probability density function $f(x) = F'(x)$ evaluated at $x = Q(u)$; in symbols, $fQ(u) = f(Q(u))$. We call $fQ(u)$ the density-quantile function.

Estimating the fQ-function is of interest for two reasons: as a way of (1) estimating the true probability density function $f(x)$, and (2) estimating approximate confidence intervals for $Q(u)$ and especially for the true median $Q(0.5)$, since

$$\tilde{Q}(0.5) \pm \{\sqrt{n}\, fQ(0.5)\}^{-1} \tag{5}$$

is an approximate 95% confidence interval for the median $Q(0.5)$.

We call $q(u) = Q'(u)$ the quantile-density function. The identity $FQ(u) = u$ implies the reciprocal relationship

$$fQ(u) \quad q(u) = 1 \quad . \tag{6}$$

Consequently, one can estimate fQ by estimating q, and one can estimate q by differences of Q. Thus we may write (using $\doteq$ to denote approximate equality)

$$\frac{1}{fQ(0.5)} = q(0.5) \doteq \frac{Q(0.75) - Q(0.25)}{0.75 - 0.25} = 2\{Q(0.75) - Q(0.25)\} . \tag{7}$$

Symmetric differences (about 0.5) of Q provide measures of scale. Define, for $0 \le p \le 0.5$, $S(p) = Q(1 - p) - Q(p)$ to be the p-spread (or p-range), and $\tilde{S}(p) = \tilde{Q}(1 - p) - \tilde{Q}(p)$ to be the sample p-spread. When $p = 0.25$, one calls $Q(0.75)$ and $Q(0.25)$ the quartiles, $S(0.25) = Q(0.75) - Q(0.25)$ the quartile-spread and $\tilde{S}(0.25) = \tilde{Q}(0.75) - \tilde{Q}(0.25)$ the sample quartile-spread.

One can conclude that the median $Q(0.5)$ has a non-parametric estimator given by $\tilde{Q}(0.5)$, and an approximate 95% confidence interval given by

$$\tilde{Q}(0.5) \pm 2\tilde{S}(0.25)/\sqrt{n} \quad . \tag{8}$$

A use of a confidence interval of this kind for the median is discussed by McGill, Tukey, and Larsen (1978).

4. LOCATION AND SCALE PARAMETER ESTIMATION FOR SYMMETRIC PROBABILITY LAWS.

One of the points which this paper would like to make is that measures of location and scale of a data sample are interpretable only if they are probability-based, in the sense that they are estimators of characteristics of the true quantile function of the random variable X.

We use μ and σ to denote measures of location and scale respectively. When μ and σ represent median and inter-quartile range, $\mu = Q(0.5)$ and $\sigma = Q(0.75) - Q(0.25)$. When μ and σ^2 represent mean and variance, they can be expressed in terms of Q by

$$\mu = \int_0^1 Q(u)\, du \quad , \qquad \sigma^2 = \int_0^1 \{Q(u) - \mu\}^2\, du \quad .$$

These formulas follow immediately from the basic fact that X is identically distributed as $Q(U)$ where U is uniformly distributed on the interval $[0,1]$.

When μ and σ^2 represent mean and variance, fully non-parametric estimators of μ and σ^2 are

$$\tilde{\mu} = \int_0^1 \tilde{Q}(u)\, du \quad , \quad \tilde{\sigma}^2 = \int_0^1 \{\tilde{Q}(u) - \tilde{\mu}\}^2\, du \quad ,$$

which are essentially the sample mean and the sample variance. Robust statistical inference starts with the warning that these estimators can be very undesirable in practice [see Mosteller and Tukey (1977), p. 203].

To interpret location and scale parameters μ and σ, it is customary to start with a model for the probability density function $f(x)$ of the data:

$$f(x) = \frac{1}{\sigma} f_0\left(\frac{x - \mu}{\sigma}\right) \tag{1}$$

where $f_0(x)$ is a known probability density function. The model (1) can be stated equivalently in terms of the true quantile function $Q(u)$:

$$Q(u) = \mu + \sigma\, Q_0(u) \tag{2}$$

where $Q_0(u)$ is a known quantile function corresponding to $f_0(x)$.

An L-estimator $\hat{\mu}$ of a location parameter is a linear combination of order statistics $X_{(1)} < \ldots < X_{(n)}$, which we write in the form

$$\hat{\mu} = \int_0^1 \tilde{Q}(u)\, W(u)\, du \tag{3}$$

for a suitable weight function $W(u)$. Asymptotically efficient L-estimators of μ and σ in the model $Q(u) = \mu + \sigma\, Q_0(u)$, when f_0 is a <u>known</u> symmetric density, are given by [see Parzen (1978), and summary in the next section].

$$\hat{\mu} = \int_0^1 \tilde{Q}(u)\, W_\mu(u)\, du \div \int_0^1 W_\mu(u)\, du \tag{4}$$

$$\hat{\sigma} = \int_0^1 \tilde{Q}(u)\, W_\sigma(u)\, du \div \int_0^1 Q_0(u)\, W_\sigma(u)\, du \tag{5}$$

where

$$W_\mu(u) = f_0Q_0(u)\, J_0'(u) = \frac{J_0'(u)}{Q_0'(u)} \tag{6}$$

$$W_\sigma(u) = J_0(u) + Q_0(u)\, W_\mu(u) \quad . \tag{7}$$

$f_0Q_0(u)$ is the density-quantile function corresponding to Q_0, and $J_0(u)$ is its score function defined by

$$J_0(u) = -(f_0Q_0)'(u) = \frac{-f_0'Q_0(u)}{f_0Q_0(u)} = \psi(Q_0(u)) \ , \tag{8}$$

where

$$\psi(x) = -\frac{f_0'(x)}{f_0(x)} = \frac{-d}{dx}\log f_0(x) \quad . \tag{9}$$

An L-estimator forms a weighted average of order statistics in which the weights depend on the <u>ranks</u> u. It is of interest to express the weights as a function of $Q_0(u)$, which is the <u>size</u> of the order statistics:

$$W_\mu(u) = \psi'(Q_0(u)) \tag{10}$$

$$W_\sigma(u) = \{\psi(x) + x\psi'(x)\}\Big|_{x = Q_0(u)} \quad . \tag{11}$$

An important conclusion is that an L-estimator $\hat{\mu}$ of μ for a symmetric probability density can be expressed

$$\hat{\mu} = \frac{\int_0^1 w(Q_0(u))\, \tilde{Q}(u)\, du}{\int_0^1 w(Q_0(u))\, du} \tag{12}$$

for a suitable choice of weight function $w(x)$.

We call w a <u>window</u> because in practice we will generate estimators $\hat{\mu}$ by specifying w and using equation (14) below. For a specified window w, one needs to study for what data models it provides good estimators.

As written in (12) the estimator will require a model to be specified because Q_0 appears. To obtain estimators $\hat{\mu}$ and $\hat{\sigma}$ which are completely model-free, data analytic, and hopefully robust one uses an iterative process in which estimators of one iteration are used to generate estimators of the next iteration.

Let μ^* and σ^* denote estimators of μ and σ. Define

$$\tilde{Q}_0^*(u) = \frac{\tilde{Q}(u) - \mu^*}{\sigma^*} \quad ; \tag{13}$$

it can be regarded as a raw estimator of the unknown quantile function $Q_0(u)$. We propose the following definitions for the estimators of the next iteration:

$$\hat{\mu} = \frac{\int_0^1 \tilde{Q}(u)\, w(\tilde{Q}_0^*(u))\, du}{\int_0^1 w(\tilde{Q}_0^*(u))\, du} \tag{14}$$

$$\hat{\sigma}^2 = \frac{\int_0^1 \{\tilde{Q}(u) - \hat{\mu}\}^2\, w(\tilde{Q}_0^*(u))\, du}{\int_0^1 \{\tilde{Q}_0^*(u)\}^2\, w(\tilde{Q}_0^*(u))\, du} \quad . \tag{15}$$

We call $\hat{\mu}$ and $\hat{\sigma}$ iteratively reweighted estimators of μ and σ.

These estimators can also be motivated from the point of view of maximum likelihood estimation (which leads to M-estimates in the literature of robust estimation). Define $L(\mu, \sigma)$ to be $(1/n)$ times the log - likelihood of the sample $X_1, \ldots, X_n$; it is given by

$$L(\mu, \sigma) = -\log \sigma + \frac{1}{n} \sum_{i=1}^{n} \log f_0 \left(\frac{X_i - \mu}{\sigma} \right) .$$

One can express likelihood in terms of quantile functions:

$$L(\mu, \sigma) = -\log \sigma + \int_0^1 \log f_0 \left(\frac{\tilde{Q}(u) - \mu}{\sigma} \right) du \quad . \tag{16}$$

For ease of writing we introduce the notation

$$\tilde{Q}_0(u) = \frac{\tilde{Q}(u) - \mu}{\sigma} \quad . \tag{17}$$

The maximum likelihood estimators $\hat{\mu}$ and $\hat{\sigma}$ satisfy the log likelihood-derivative equations:

$$\frac{\partial}{\partial \mu} L(\hat{\mu}, \hat{\sigma}) = 0 \quad , \quad \frac{\partial}{\partial \sigma} L(\hat{\mu}, \hat{\sigma}) = 0 \quad . \tag{18}$$

To compactly write formulas for these derivatives, define

$$w(x) = \frac{1}{x}\,\psi(x) \tag{19}$$

where $\psi(x)$ is defined by (9). Then

$$\begin{aligned}\frac{\partial}{\partial\mu} L(\mu, \sigma) &= \frac{1}{\sigma}\int_0^1 \psi(\tilde{Q}_0(u))\, du \\ &= \frac{1}{\sigma^2}\int_0^1 \{\tilde{Q}(u) - \mu\}\, w(\tilde{Q}_0(u))\, du\end{aligned} \tag{20}$$

$$\begin{aligned}\frac{\partial}{\partial\sigma} L(\mu, \sigma) &= -\frac{1}{\sigma} + \frac{1}{\sigma^2}\int_0^1 \psi(\tilde{Q}_0(u))\,\{\tilde{Q}(u) - \mu\}\, du \\ &= -\frac{1}{\sigma} + \frac{1}{\sigma^3}\int_0^1 w(\tilde{Q}_0(u))\,\{\tilde{Q}(u) - \mu\}^2\, du \quad .\end{aligned} \tag{21}$$

In the normal case, $\psi(x) = x$, $w(x) = 1$ and $\hat{\mu}$ and $\hat{\sigma}^2$ are equal to the sample mean and variance respectively.

To obtain estimators $\hat{\mu}$ and $\hat{\sigma}$ without specifying $f_0(x)$, one studies the iteratively reweighted estimators of μ and σ^2 defined by (14) and (15). They can be regarded as "approximate" solutions of the log-likelihood derivative equations, given prior estimates μ^* and σ^* .

When we are concerned with forming estimators of location and scale which are satisfactory for long-tailed distributions it is natural to choose weight functions $w(x)$ corresponding to Students' t-distribution with m degrees of freedom,

$$f_0(x) = \frac{1}{\sqrt{m\pi}}\;\frac{\Gamma(\frac{m+1}{2})}{\Gamma(\frac{m}{2})}\left(1 + \frac{x^2}{m}\right)^{-(m+1)/2} , \tag{22}$$

for which

$$w(x) = -\frac{1}{x}\,(\log f_0(x))' = \frac{m+1}{m}\;\frac{1}{1 + \frac{1}{m}x^2} \quad . \tag{23}$$

To completely specify the window $w(x)$, one must specify a value for m (which one could call the "trimming width" of the window). The more normal the data is believed to be, the larger should m be chosen (say,

$m = 25$). The more Cauchy-distributed the data is believed to be, the closer to 1 should m be chosen (say, $m = 4$). In practice, one might try both values of m, and compare the results. The constant m could also be estimated adaptively to yield "self-tuning" robust estimators of location and scale. We call $w(x)$ the Student window.

A window recommended by Tukey [see Mosteller and Tukey (1977), p. 205] is the bisquare window:

$$w_{\text{Bisquare}}(x) = (1 - (\frac{x}{c})^2)_+^2 \tag{24}$$

where c is a suitably chosen constant. Tukey recommends that c be taken to be 6 or 4 when x is measured in units of σ. It seems likely that the choice of c should reflect one's beliefs about the long-tailed character of the data.

Windows with noteworthy properties are Huber's favorite choices [Huber (1977), p. 13]:

$$\begin{aligned} w_H(x) &= 1 \qquad \text{if } |x| \le k \\ &= \frac{k}{|x|} \qquad \text{if } |x| \ge k \end{aligned} \tag{25}$$

which depends upon a parameter k. These windows are optimal for the normal-center and exponential-tailed density

$$\begin{aligned} f_0(x) &= c\, e^{-\frac{1}{2}x^2} \quad , \quad |x| \le k \\ &= c\, e^{\frac{1}{2}k^2} e^{-k|x|} \quad , \quad |x| \ge k \end{aligned} \tag{26}$$

where c is a constant chosen to make the density integrate to 1 .

For Student's t-distribution with m degrees of freedom, the weight functions of L-estimators of μ and σ are given by $W_\mu(u) = w_\mu(Q_0(u))$, $W_\sigma(u) = w_\sigma(Q_0(u))$ with

$$w_\mu(x) = \frac{m+1}{m} \frac{1 - (x^2/m)}{[1 + (x^2/m)]^2} \quad , \quad w_\sigma(x) = \frac{m+1}{m} \frac{2x}{[1 + (x^2/m)]} \ . \tag{27}$$

These windows deserve further investigation. However, they appear to support the recommendation that robust estimators of location and scale may be obtained from iterative estimators generated from the Student window (23) with possibly iteratively chosen values of m. For the Student window, formulas for iteratively reweighted estimators $\hat{\mu}$ and $\hat{\sigma}$ are explicitly given by

$$\hat{\mu} = \frac{\int_0^1 \tilde{Q}(u) \{1 + \frac{1}{m}(\frac{\tilde{Q}(u) - \mu^*}{\sigma^*})^2\}^{-1} du}{\int_0^1 \{1 + \frac{1}{m}(\frac{\tilde{Q}(u) - \mu^*}{\sigma^*})^2\}^{-1} du} \tag{28}$$

$$\hat{\sigma}^2 = \frac{\int_0^1 \{\tilde{Q}(u) - \hat{\mu}\}^2 \{1 + \frac{1}{m}(\frac{\tilde{Q}(u) - \mu^*}{\sigma^*})^2\}^{-1} du}{\int_0^1 \{\frac{\tilde{Q}(u) - \mu^*}{\sigma^*}\}^2 \{1 + \frac{1}{m}(\frac{\tilde{Q}(u) - \mu^*}{\sigma^*})^2\}^{-1} du} . \tag{29}$$

5. LOCATION AND SCALE PARAMETER ESTIMATION FOR CONTAMINATED NORMAL DATA.

One can form estimators, denoted $\hat{\mu}_{p,q}$ and $\hat{\sigma}_{p,q}$, which are the most efficient estimators μ and σ using the sample quantile function $\tilde{Q}(u)$, $p \le u \le q$, over a sub-interval; this is equivalent to using a restricted set of order statistics $X_{(np)}, \ldots, X_{(nq)}$ or a trimmed sample. A compact derivation of such estimators is given by Parzen (1978) who gives the representation

$$\begin{bmatrix} \hat{\mu}_{p,q} \\ \hat{\sigma}_{p,q} \end{bmatrix} = \begin{bmatrix} I_{\mu\mu} & I_{\mu\sigma} \\ I_{\mu\sigma} & I_{\sigma\sigma} \end{bmatrix}^{-1} \begin{bmatrix} T_{\mu,p,q} \\ T_{\sigma,p,q} \end{bmatrix} \tag{1}$$

where

$$T_{\mu,p,q} = \int_p^q W_\mu(u)\, \tilde{Q}(u)\, du + \tilde{Q}(p)\, W_{\mu L}(p) + \tilde{Q}(q)\, W_{\mu R}(q)$$

$$T_{\sigma,p,q} = \int_p^q W_\sigma(u)\, \tilde{Q}(u)\, du + \tilde{Q}(p)\, W_{\sigma L}(p) + \tilde{Q}(q)\, W_{\sigma R}(q)$$

$$I_{\mu\mu} = \int_p^q W_\mu(u)\, du + W_{\mu L}(p) + W_{\mu R}(q) \tag{2}$$

$$I_{\mu\sigma} = \int_p^q W_\sigma(u)\,du + W_{\sigma L}(p) + W_{\sigma R}(q)$$

$$I_{\sigma\sigma} = \int_p^q W_\sigma(u)\,Q_0(u)\,du + W_{\sigma L}(p)\,Q_0(p) + W_{\sigma R}(q)\,Q_0(q) \ .$$

The weight functions are expressed in terms of the density-quantile function $f_0Q_0(u)$ and the score function $J_0(u)$ by

$$
\begin{aligned}
W_\mu(u) &= J_0'(u)\,f_0Q_0(u) \\
W_\sigma(u) &= J_0(u) + Q_0(u)\,W_\mu(u) \\
W_{\mu L}(p) &= f_0Q_0(p)\left[\frac{1}{p}\,f_0Q_0(p) + J_0(p)\right] \\
W_{\mu R}(q) &= f_0Q_0(q)\left[\frac{1}{1-q}\,f_0Q_0(p) - J_0(p)\right] \\
W_{\sigma L}(p) &= Q_0(p)\,W_{\mu L}(p) - f_0Q_0(p) \\
W_{\sigma R}(q) &= Q_0(q)\,W_{\mu R}(p) + f_0Q_0(q) \ .
\end{aligned}
\tag{3}
$$

For normally distributed data,

$$
\begin{aligned}
f_0(x) &= \phi(x) = \frac{1}{\sqrt{2\pi}}\,e^{-(1/2)x^2} \ , \quad F_0(x) = \Phi(x) = \int_{-\infty}^{x} \phi(y)\,dy \ , \\
f_0Q_0(u) &= \phi\Phi^{-1}(u) = \frac{1}{\sqrt{2\pi}}\,\exp -\frac{1}{2}\,|\Phi^{-1}(u)|^2 \ , \\
J_0(u) &= \Phi^{-1}(u) \ , \quad J_0'(u) = \{\phi\,\Phi^{-1}(u)\}^{-1} \ , \\
W_\mu(u) &= 1 \ , \quad W_\sigma(u) = 2\,\Phi^{-1}(u) \ . \\
W_{\mu L}(p) &= \phi\,\Phi^{-1}(p)\,\{\frac{1}{p}\phi\,\Phi^{-1}(p) + \Phi^{-1}(p)\} \\
W_{\sigma R}(p) &= \Phi^{-1}(p)\,W_{\mu L}(p) - \phi\Phi^{-1}(p) \ .
\end{aligned}
\tag{4}
$$

When $q = 1 - p$, $I_{\mu\sigma} = 0$,

$$I_{\mu\mu} = 1 - 2p + 2W_{\mu L}(p)$$
$$I_{\sigma\sigma} = 2\int_p^q |\Phi^{-1}(u)|^2 \, du + 2\Phi^{-1}(p)\, W_{\sigma L}(p) \; . \tag{5}$$

The estimator

$$\hat{\mu}_{p,q} = \frac{\int_p^{1-p} \tilde{Q}(u)\, du + W_{\mu L}(p)\{\tilde{Q}(p) + \tilde{Q}(1-p)\}}{1 - 2p + 2W_{\mu L}(p)} \tag{6}$$

is similar to the Winsorized mean (with trimming proportion p).

Robust Maximum Likelihood Estimation of Mean and Variance of a Normal Distribution. We may be willing to assume that our data is more normal than long-tailed, but the shape of the true distributions is deviating slightly from the assumed normal model due to "wrong" values of the data set. We propose the following exploratory data analysis for robust estimation of μ and σ from normal data with possible "outliers." We suggest the name "robust maximum likelihood estimators" for these estimators.

For selected values of p (at least $p = 0.05$, 0.25 , and 0.45) , (1) compute the estimators $\hat{\mu}_{p,1-p}$ and $\hat{\sigma}_{p,1-p}$, and (2) plot the residuals

$$\tilde{Q}(u) - \hat{Q}(u) = \tilde{Q}(u) - \hat{\mu}_{p,1-p} - \hat{\sigma}_{p,1-p}\, \Phi^{-1}(u) \quad ,$$

multiplied by $\phi\Phi^{-1}(u)$. Their values over the interval $p \le u \le 1 - p$ can be used to test the hypothesis H_0 . The residuals over the tail intervals $u \le p$ and $u \ge 1 - p$ can be used to test for the presence of "wrong values." One estimates μ and σ by those estimators corresponding to the lowest value of p for which one finds no "wrong values" over the tail intervals.

6. NONPARAMETRIC PROBABILITY LAW MODELING.

Parametric statistical inference assumes that the data to be analyzed comes from a single population whose probability law is known up to a finite number of parameters. Goodness of fit procedures test whether such an hypothesis fits the data. Robust statistical inference is concerned with parameter estimation when the data is "mildly dirty" (such as contaminated normal) or obeys an assumption such as a unimodal probability density symmetric about its mode for which location, regression, and scale parameters can still be interpreted. One can develop goodness of fit procedures for these qualitatively defined ideal models related to non-parametric density estimation procedures (which also provide goodness-of-fit tests for parametric ideal models).

To say that the true quantile function $Q(u)$ obeys the hypothesis $H_0 : Q(u) = \mu + \sigma Q_0(u)$ is to say that one can find values $\hat{\mu}$ and $\hat{\sigma}$ such that $\hat{Q}(u) = \hat{\mu} + \hat{\sigma} Q_0(u)$ fits $\tilde{Q}$. The fit of $\hat{Q}$ to $\tilde{Q}$ can be judged by displaying the quantile residuals

$$R(u) = f_0 Q_0(u) \{\tilde{Q}(u) - \hat{Q}(u)\} , \quad 0 \le u \le 1$$

where $f_0 Q_0(u) = f_0(Q_0(u))$ is the density-quantile function corresponding to Q_0. Under the null hypothesis H_0, $(\sqrt{n}/\sigma) R(u)$, $0 \le u \le 1$ is asymptotically distributed as a stochastic process $B(u)$, $0 \le u \le 1$ which is a modified Brownian Bridge process in the sense that its covariance kernel $E[B(u_1) B(u_2)]$ is not $\min(u_1, u_2) - u_1 u_2$ but is modified due to the estimation of the parameters μ and σ. One tests whether the sample path $R(u)$ looks like a sample path from such a modified Brownian Bridge process.

As shown in Parzen (1978), one can test whether the true quantile function $Q(u)$ obeys H_0, without first estimating μ and σ, by forming

$$\tilde{D}(u) = \frac{1}{\tilde{\sigma}_0} \int_0^u f_0 Q_0(t)\, d\tilde{Q}(t) , \quad 0 \le u \le 1 \quad ; \quad \tilde{\sigma}_0 = \int_0^1 f_0 Q_0(u)\, d\tilde{Q}(u) ,$$

which is an estimator of

$$D(u) = \frac{1}{\sigma_0} \int_0^u f_0 Q_0(t)\, dQ(t) , \quad 0 \le u \le 1 \quad ; \quad \sigma_0 = \int_0^1 f_0 Q_0(u)\, dQ(u) .$$

Under the null hypothesis, $D(u) = u$, and it is conjectured that $\sqrt{n}\, \{\tilde{D}(u) - u\}$, $0 \le u \le 1$ is asymptotically distributed as a Brownian Bridge Stochastic process. By suitably choosing the null hypothesis H_0

and the standard density-quantile function f_0Q_0, one can test the goodness of fit of any specified probability law (normal, exponential, Weibull, Cauchy, etc.) to the data.

Parzen (1978) discusses estimators $\hat{d}(u)$ of

$$d(u) = D'(u) = \frac{1}{\sigma_0} \frac{f_0Q_0(u)}{fQ(u)}$$

which can be used to form estimators of $fQ(u)$.

Estimators of $fQ(u)$ in fact yield a formula for Q_0, and also generate $W_\mu(u)$ and $W_\sigma(u)$ which provide formulas for $\hat{\mu}$ and $\hat{\sigma}$ as linear combinations of order statistics; these estimators of location and scale yield a representation $Q(u) = \hat{\mu} + \hat{\sigma} Q_0(u)$ which is "adaptive" (in the sense that the null hypothesis quantile function Q_0 has been estimated from the data).

Estimating the density-quantile function $fQ(u)$ provides an analytical approach for checking data (or regression residuals) for unimodality and symmetry, which are the assumptions underlying certain applications of robust statistical procedures. Another approach is the Quantile-Box Plot which is a graphical display of the quantile function $\tilde{Q}(u)$ as a function on the unit interval $0 \leq u \leq 1$, on which is superimposed boxes representing respectively the 1/4 and 3/4 percentile, the 1/8 and 7/8 percentiles, and the 1/16 and 15/16 percentiles (see Figures). It is a variation of the Box-Plot introduced by Tukey (1977).

Quantile-Box Plots enable the investigator to detect "non-ideal" aspects of the data batches by testing the data for normality by tests which determine the directions in which data fails to be normal, such as (1) long-tailed distribution, (2) outliers, (3) bimodal distribution, (4) non-symmetric distribution [see Parzen (1978)].

A Quantile-Box Plot of a data batch $X_1, \ldots, X_n$ enables one to classify it into one of four categories:

I. Continuous, unimodal, symmetric.

II. Continuous, unimodal, non-symmetric.

III. Continuous, bimodal.

IV. Discrete.

Data that is classified as continuous, unimodal, and symmetric can be regarded as having passed a goodness-of-fit test for robust procedures based on these assumptions.

Data that is classified as continuous, unimodal, but non-symmetric could be transformed to symmetry by a suitable transformation (or re-expression) of the data.

Data that is classified as bimodal may consist of two populations rather than be considered a contaminated main population. For such data, robust estimation of location and scale estimators would be a pointless exercise as the results would have no interpretation.

The question of how to treat data classified as discrete requires further study.

7. ROBUST REGRESSION.

The efficient estimation of location and scale parameters can be formulated as a problem of weighted regression of the sample quantile function, with weights a function of $Q_0(u)$; this may explain the ability of robust estimators to achieve approximately efficient estimators of μ and σ by iterating ordinary regression calculations with windows that only qualitatively depend on the true distributions.

The iteratively reweighted estimators $\hat{\mu}$, defined by equation (14) of section 4, is numerically equivalent to the solution to the problem of estimating μ in the following weighted least squares linear regression problem:

$$X_j = \mu + \epsilon_j \tag{1}$$

where ϵ_j are independent normal with mean 0 and variance satisfying

$$\mathrm{Var}(\epsilon_j) = \sigma^2 \frac{1}{w_j} , \quad w_j = w(\epsilon_j^*) , \quad \epsilon_j^* = \frac{X_j - \mu^*}{\sigma^*} . \tag{2}$$

A regression model (linearly relating a variable Y to variables $X_1, \ldots, X_k$) can be written

$$Y_j = \beta_1 X_{1j} + \ldots + \beta_k X_{kj} + \epsilon_j , \quad j = 1, \ldots, n , \tag{3}$$

where $\{\epsilon_j\}$ are independent random variables with quantile function known up to a parameter σ

$$Q_\epsilon(u) = \sigma Q_0(u) \quad . \tag{4}$$

The normal case corresponds to $Q_0(u) = \Phi^{-1}(u)$.

To robustly estimate the coefficients $\beta_1, \ldots, \beta_k$, σ one first obtains initial estimators by ordinary least squares linear regression. If $\beta_1^*, \ldots, \beta_k^*$ are the estimators after a number of iterations, to obtain the next iteration (denoted $\hat{\beta}_1, \ldots, \hat{\beta}_k$) form the residuals

$$\epsilon_j^* = (Y_j - \beta_1^* X_{1j} + \ldots + \beta_k^* X_{kj}) \div \sigma_j^* \quad , \tag{5}$$

and compute estimators $\hat{\beta}_1, \ldots, \hat{\beta}_k$, $\hat{\sigma}^2$ as the weighted least squares linear regression estimators under the assumption that ϵ_j have variances defined by (2) . This process of iteration can be shown to yield robust estimators when the windows are chosen appropriately (see Huber (1977), p. 38, Algorithm W).

Point III of this paper consists of the following proposals:

(1) At each iteration, the long-tailed character of the residuals ϵ_j^* should be examined, using Quantile-Box plots and diagnostics.

(2) There is a danger that researchers may regard robust regression procedures as a routine solution to the problem of modeling relations between variables without first studying the distribution of each variable. This may be true when there is a theoretical basis for the linear relations being estimated. Otherwise, one is operating in the exploratory data analysis mode in which one must first determine ways of re-expressing the data before fitting linear regression models.

(3) One might consider non-parametric non-linear regression of Y on $X_1, \ldots, X_k$ using the density-quantile approach described by Parzen (1977) (which is being extended by J. P. Carmichael and has been applied to time series analysis by M. Pagano). It may provide means of checking whether robust estimation of variances and correlations is provided by robust estimation of linear regression coefficients.

8. EXAMPLES OF QUANTILE-BOX PLOTS.

Quantile-Box Plots can be used to compare several data samples of a random variable X under different experimental conditions. We illustrate this for Tippett's Warp Break Data quoted by Tukey (1977), p. 82 [for an example of box-plots of these data see McGill, Tukey, Larsen (1978)].

Quantile-Box Plots can be used to find appropriate transformations of the component variables of a data vector. We illustrate this application using fossil data (from yellow Limestone formation of northwestern Jamaica) quoted by Chernoff (1973). The plots show that: variables 2 and 6 have zeroes in fQ (rises in Q); variables 3 and 4 have probability masses (flat stretches in Q); variables 1 and 5 are candidates for re-expression (logarithm for 1 , square root for 5). It turns out that a scatter plot of variables 2 and 5 suffice to classify the observations into three distinct clusters.

REFERENCES

1. Chernoff, H. (1973). "The use of faces to represent points in k-dimensional space graphically," Journal of the American Statistical Assn, 68, 361.
2. Huber, P. (1977). Robust Statistical Procedures. Regional Conference Series in Applied Mathematics 27. SIAM: Philadelphia.
3. McGill, R., Tukey, J. W., Larsen, W. A. (1978). "Variations of box plots," American Statistician, 32, 12-16.
4. Mosteller, F. and Tukey, J. W. (1977). Data Analysis and Regression, Addison Wesley: Reading, Mass.
5. Parzen, E. (1977). "Nonparametric statistical data science (a unified approach based on density estimation and testing for "White Noise"). Technical Report No. 47, Statistical Science Division, State University of New York at Buffalo.
6. ________ (1978). "Nonparametric statistical data modeling," Journal of the American Statistical Assn. December issue.
7. Tukey, J. W. (1977). Exploratory Data Analysis. Addison Wesley: Reading, Mass.

Research supported by the Army Research Office (Grant DA AG29-76-0239).

Institute of Statistics
Texas A & M University

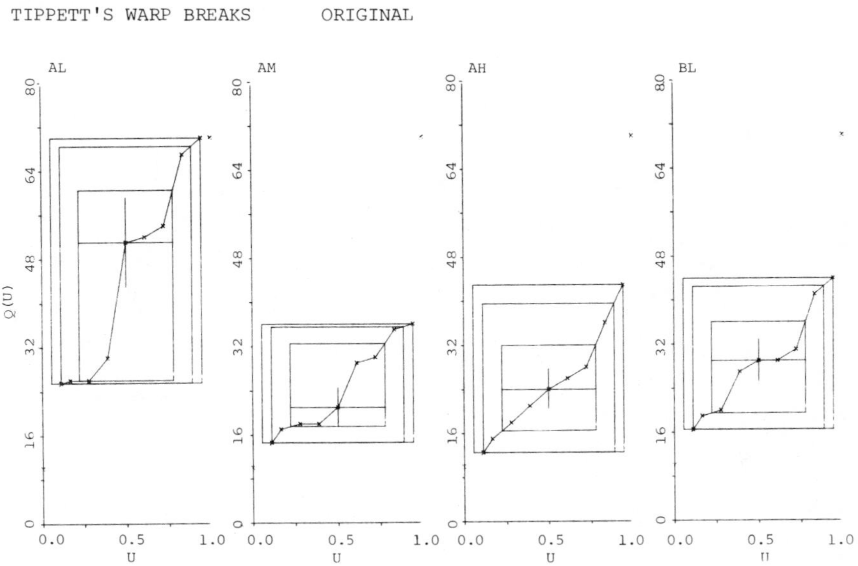
TIPPETT'S WARP BREAKS
ORIGINAL
AL
AM
AH
BL
Q(U)
U

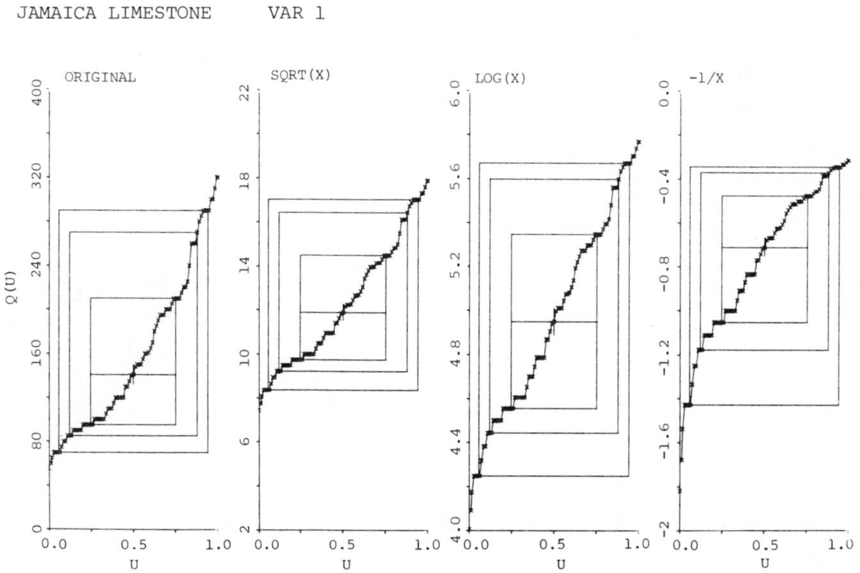
JAMAICA LIMESTONE
VAR 1
ORIGINAL
SQRT(X)
LOG(X)
-1/X
Q(U)
U

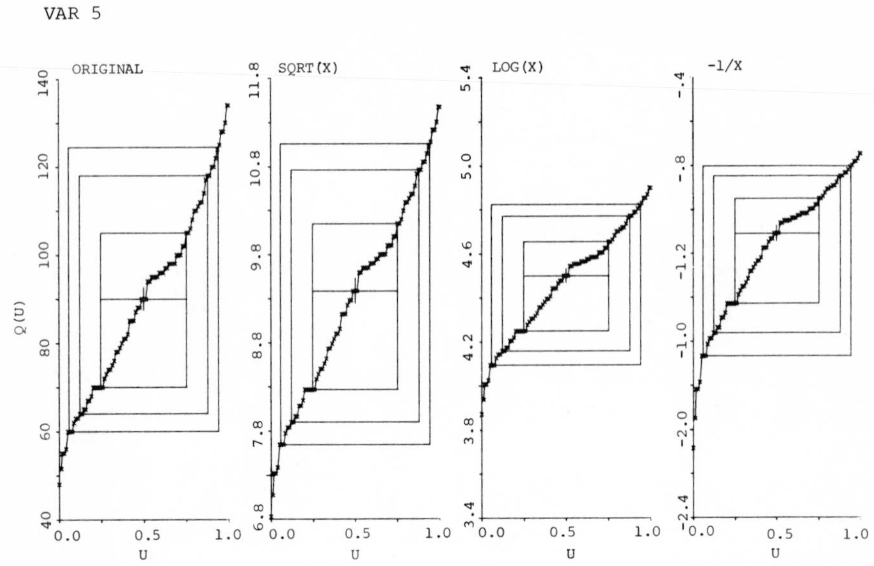
VAR 5
ORIGINAL
SQRT(X)
LOG(X)
-1/X
Q(U)
U

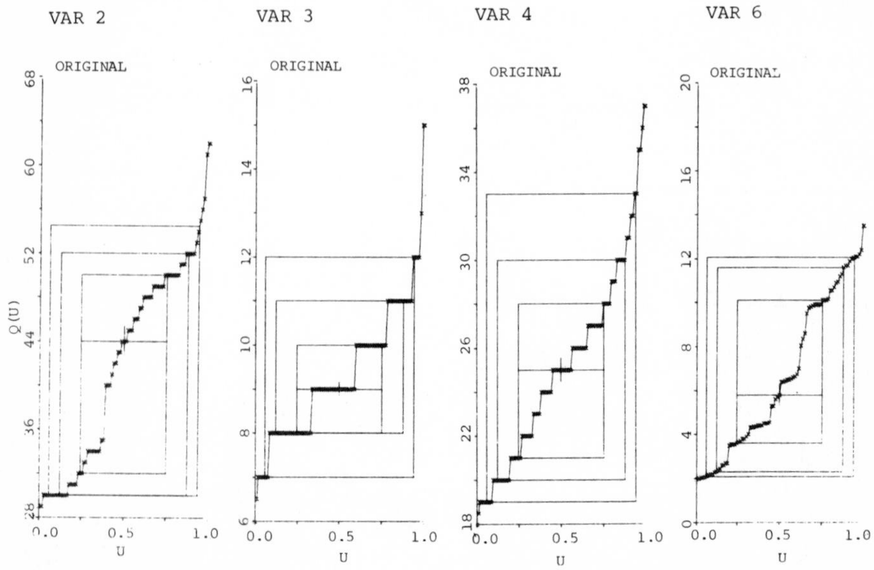
VAR 2
VAR 3
VAR 4
VAR 6
ORIGINAL
Q(U)
U

Robust Inference—The Fisherian Approach

Graham N. Wilkinson

1. INTRODUCTION

The last two decades have seen a resurgence of what Tukey (1977) has termed 'exploratory data analysis', in part a rebellion against widespread misapplication of standard classical methods of analysis assuming approximate Normality of the sampling variation. Most of the related research has concentrated on the development of 'robust' statistical methods, so-named for the property of being insensitive to outliers and more generally to possible variations in the underlying form of sampling distribution. The term 'robust' is due to G. E. P. Box (1953).

Useful ideas have emerged from this research, and it is therefore not in denigration of them that I wish here to raise some questions about the principles on which most robustness research has been based, particularly with regard to the terminal phase in statistical inference of determining confidence intervals or, more fully, confidence distributions for the parameters of interest. Here I shall suggest that the Fisherian concept of conditional parametric inference will prove to be the soundest theoretical basis for a properly adaptive approach to statistical inference, allowing the data to speak for themselves as fully as possible; some computational results in support of this view will be presented.

ISBN 0-12-438150-2

It may be noted that the conditional parametric approach is also inherent in the Bayesian approach to robustness. The present paper relates in part to a Bayesian discussion of robustness in Box and Tiao (1962).

As in so many other fields of statistics it was R. A. Fisher (1935) who produced the germinal idea of robustness, in his book The Design of Experiments, to which we return below. On the question of outliers and use of the arithmetic mean, the following quotation from his first fundamental paper on the foundations of theoretical statistics (Fisher, 1922) shows Fisher's early awareness of the issues:

> "This example [relating to the Cauchy distribution] serves also to illustrate the practical difficulty which observers often find, that a few extreme observations appear to dominate the value of the mean. In these cases the rejection of extreme values is often advocated, and it may often happen that gross errors are thus rejected. As a statistical measure, however, the rejection of observations is too crude to be defended: and unless there are other reasons for rejection than mere divergence from the majority, it would be more philosophical to accept these extreme values, not as gross errors, but as indications that the distribution of errors is not normal. As we shall show, the only Pearsonian curve for which the mean is the best statistic for locating the curve, is the normal or Gaussian curve of errors. If the curve is not of this form the mean is not necessarily of any value whatever. The determination of the true curves for different types of work is therefore of great practical importance ... ."

In general there is no doubt that Fisher regarded the proper quantitative reduction and interpretation of any body of observations as contingent on the formulation of appropriate quantitative models for the data. For in developing his general theory of statistical estimation and inference he saw clearly the need to separate relevant from irrelevant information and to discover further, in the relevant data, the ancillary information on which conditioning of the sampling distribution of the informative statistics is necessary to recover fully all the relevant information on the parameters of interest, as he showed in Fisher (1934); this inferentially essential separation can be accomplished only with reference to an appropriately formulated model or class of models.

In The Design of Experiments, Chapter III, Fisher raises for the first time the issue of robustness in terms of the sensitivity of the t-test to the underlying assumption of Normality. Having illustrated the use of the t-test on 15 differences in height between paired cross- and self-fertilized plants drawn from data of Darwin (1876), he then considers the wider null hypothesis that the two types of plants have the same true but unspecified distribution of height, and develops the now-famous Randomization test, noting that (if Darwin's experiment had been properly randomized), any of the 15 observed differences in height on this hypothesis might equally have occurred with opposite sign under a different randomization, and therefore that the mean (Fisher refers to sum) of the observed differences in height could be judged in relation to all the 2^{15} possible means obtainable with reversals of sign, all equally likely under randomization. This 'distribution-free' test gives surprisingly close agreement with the t-test on Darwin's data, producing an upper-tail probability of 2.634% compared with 2.485%. Fisher leaves unsaid at this point the implication that the t-test is therefore robust against departures from Normality but is more explicit later, in Chapter IV:

> "The z test [or F test, in Analysis of Variance] ... is even less affected than the t-test by such deviations from normality as are met with in practice."

Criterion Robustness, as it is often termed, is the lynch-pin of most subsequent robustness research, yet there can be no doubt in retrospect that as a criterion of inferential validity it is fundamentally unsound, as Box and Tiao (1962) noted in discussing the same Darwin data. Fisher undoubtedly was misled by the remarkable agreement he found between the results of the t-test and the randomization test and failed to perceive, apparently, an important element of arbitrariness in the reasoning. For had he substituted some other location estimator than the arithmetic mean, say the maximum-likelihood estimator of location for some long-tailed distribution, the same randomization argument would have led to quite a different probability value for the Darwin data, and indeed a host of different probabilities could be obtained according to the estimator chosen.

The chief defect of most of the current robustness criteria is, in the context of statistical inference, that their reliance on marginal frequency properties of a criterion can be in serious conflict with the central, conditioning principle of Fisherian theory. Distribution-free (or non-parametric) tests also fail to utilize properly the information provided by the data themselves on distributional form, a defect which Fisher himself recognized. The point is discussed in a new Section 21.1 added to Chapter III in the 7th edition (1960) of _The Design of Experiments_.

The main purpose of the present paper is to illustrate the central role for Fisherian conditional parametric inference theory in robustness studies by extending the kind of comparison made by Fisher above to confidence distributions, utilizing an important generalization due to Hartigan (1969) of the randomization argument. After some preliminary remarks in Section 2 I outline briefly in Section 3 the basic principles of statistical inference on which the Fisherian theory is based, and the conditional fiducial-confidence distributions to which they lead. In Section 4 the essential result of Hartigan (1969) on distribution-free marginal confidence distributions is described, and then in Section 5, Fisherian conditional confidence distributions are compared numerically with Hartigan confidence distributions generated by the corresponding maximum-likelihood estimators of location for an exponential-power family of distributions considered by Diananda (1949), Box and Tiao (1962). Some final conclusions are presented in Section 6.

There is indeed a novel robustness property to be deduced from these numerical studies, namely that, if the sample is close to Normal <u>in appearance</u>, the conditional confidence distribution for location is almost independent of the sampling distribution assumed, and is of the classical Normal form based on t_n. There is thus a much more powerful <u>conditional</u> central limit theory to be formulated for small samples.

2. SOME PRELIMINARY REMARKS

In reasoning about scientific theories one should recognize first, as Einstein did, that any scientific principle or theory can be objectively valid only as a <u>representation of observable reality</u>, and that its validity extends no further on current evidence than some limited <u>domain of application</u> over which no conflict with known facts arises.

Likewise the only objectively valid representation of an unknown conceptual 'constant' of Nature is at best as a random variable with a probability distribution of possible values determined by the finite totality of currently available measurements of it, direct or indirect. Fisher recognized this clearly and his fiducial argument (Fisher, 1930) for deriving probability distributions of unknown parameters is really the pinnacle of his achievements in the field of statistical inference, even though he failed to elucidate its logic completely. The logical basis of fiducial probability is now known (Wilkinson, 1977), see Section 3.

A theory of statistical inference differs from purely deductive forms of scientific theory in that it produces, not statements of fact, but uncertain inferences derived by logical inversion from known evidence. It is in its highest sense a metatheory for reasoning about scientific theories of the real world in the face of uncertainties engendered by observation of it. What has not been sufficiently understood is that in relation to the supreme metatheory of logic it stands like any other practical theory, built on extra-logical principles which are the empirical 'facts' to which deductive logic may be applied. These principles are so circumscribing that in a domain of exact parametric inference they allow no latitude at all in the reasoning of statistical inferences at a fundamental level; and very little latitude in a broader domain admitting approximations to exact knowledge.

A crucial distinction must be made between Statistical Inference and Decision Theory, which have logically distinct domains of application (see Fisher (1956) and Cox (1958)). Statistical Inference in the practical sense is the science of interpreting observational evidence to produce uncertain inferences which reveal as fully as possible the meaning of the evidence in terms of laws and parameters describing the real world. This requirement is indeed sufficient, if properly formulated, to determine the inferences. There is no decision space, no loss function in relation to a specified sequence of real contingencies, but simply observational knowledge which engenders a degree of uncertainty in its interpretation, independent of the ultimate utility of such knowledge, which cannot be known in advance. Decision Theory, on the other hand, is a purely deductive theory concerned with optimal operating strategies and their factual properties in a well-specified physical context such as Quality Control or Business Management, in which the additional ingredients of a decision space, loss function etc. are specified. No attempt is made to learn all that Nature has to say, but only specific factual answers to specific questions. In effect we are dealing with a restricted form of inference in which Nature is forbidden from expressing itself fully. The answers, thus distorted by the restricted nature of the questions, may very well be appropriate for the purposes intended, but to suppose such answers singly or collectively constitute the substance of Statistical Inference is a major error, well illustrated by the serious distortions of evidence which James-Stein shrinkage estimation can induce, for instance.

3. BASIC PRINCIPLES OF STATISTICAL INFERENCE

I shall now outline briefly what I take to be unquestionable basic principles of statistical inference and show how these lead uniquely to Fisherian fiducial inference in a domain of exact non-Bayesian parametric inference. I shall then discuss the extension of the theory to the wider domain of approximate parametric inference, including the area of interest in robustness research.

3.1. The Fundamental Principle of Statistical Inference

The following, more briefly termed the Relevance Principle, is unquestionably the fundamental principle of uncertain inference, and has been known as such for centuries in British-based systems of administering Law:

Relevance Principle:

> 'Valid uncertain inference requires that all relevant evidence be properly utilized, and all irrelevant or spurious evidence excluded in the reasoning of it.'

Fisher often stressed the first part of this principle but not so explicitly the second part, which is equally important. (A familiar instance in Law is the instruction given to juries to exclude evidence ruled inadmissible or irrelevant from influencing their judgement in any whatever.)

The principle is characteristic of uncertain inference, and arises because an uncertain inference is not a statement of fact (either right or wrong) but may vary in its expressed uncertainty according to the evidence taken into account.

It must, of course, be applied in conjunction with other principles which define relevant and irrelevant evidence in various contents:

3.2. Relevant and Irrelevant Evidence

In a wide domain of exact non-Bayesian parametric inference, reduction of evidence to fully relevant form is effected by Fisher's Sufficiency and Conditioning principles, in conjunction with a third, Ordering principle. The rare case of objective Bayesian inference is covered by a Likelihood Principle. In what follows, the symbols t, z denote observables, a subscript a indicates their actual values, θ denotes possible values of the actual but unknown parameters θ_a, and $\{dF(\ ;\theta)\}$ a specified family of expected frequency distributions. The term probability will be reserved exclusively for use in the inferential sense to be described, and not as a synonym for expected frequency. To avoid discussing subtleties here the principles will be stated in the simplest form for regular cases:

Sufficiency Principle:

'If $[t_a, z_a; \{dF(t,z;\theta)\}]$ represents all the evidence pertaining to θ_a, and

$$dF(t,z;\theta) = dF(t;\theta)\cdot dF(z) \qquad (1)$$

then $[t_a; \{dF(t;\theta)\}]$ is a Sufficient representation of the evidence and $[z_a; dF(z)]$ is Irrelevant.'

Since the Relevance Principle requires all irrelevant information to be excluded, the Sufficiency Principle must be applied recursively to produce Minimal Sufficient evidence.

After this first stage of reduction (which may require additionally a contraction of the sample space of t, see Wilkinson, 1977), further reduction may be effected by conditioning on ancillary statistics, i.e., functions of the minimal sufficient statistic which are distributed independently of θ:

Conditioning Principle:

'If $[t_a, z_a; \{dF(t,z;\theta)\}]$ represents Minimal Sufficient evidence about θ_a, and

$$dF(t,z;\theta) = dF(t|z;\theta)\cdot dF(z) \ , \tag{2}$$

then $[t_a; \{dF(t|z_a;\theta)\}]$ is a Conditioned Minimal Sufficient representation of the evidence, and $[z_a; dF(z)]$ becomes Irrelevant (after conditioning).'

Note in fact that the conditional distribution function $F(t|z;\theta)$ is statistically independent of the ancillaries z, and so is carrying all the necessary information to relate θ_a to t_a, z_a. As before, the principle must be applied recursively to produce a Fully Conditioned Minimal Sufficient representation of the evidence. Fisher (1934) showed that such conditioning is necessary to recover fully the Fisher Information of the evidence in the distribution assigned to the informative statistic t.

It may seem that the Sufficiency Principle is merely a special case of the Conditioning Principle, but a counter-example of Dawid discussed in Wilkinson (1977) shows that failure to first apply the Sufficiency Principle fully before conditioning can lead to a loss of information, through conditioning on improper ancillaries.

In the rare case of Objective Bayesian evidence, where additionally it is known that θ_a was produced by a physical random process with frequency distribution $dF(\theta)$, the following likelihood principle applies:

Bayesian Likelihood Principle:

'If $[t_a; dF(\theta), \{dF(t|\theta)\}]$ represents all the evidence pertaining to θ_a, then the posterior distribution $dF(\theta|t_a)$ is the Fully Relevant representation of the evidence, depending only on $dF(\theta)$ and the realized likelihood $L(\theta;t_a)$ determined from $dF(t|\theta)$ by t_a. The sampling variation of $L(\theta;t)$ is Irrelevant.'

Note that in any other case, elimination of the sampling variation of $L(\theta;t)$ from the evidence contravenes the Relevance Principle. Non-objective Bayesian inference theories are thus fundamentally invalid in my view, though possibly still operationally useful in certain contexts.

Consideration of the Ordering Principle will be deferred until the scientific concept of inferential probability has been discussed.

3.3. Inferential Probability

In the sense in which most practising scientists understand it, Inferential Probability in an uncertain inference is the measure of certainty attaching to a proposition about the real world which is either true or false, but whose truth is currently unknown. A typical proposition might be 'It will rain here tomorrow' or '$2.1 < \mu < 5.3$' and the probability might be 1/3 or 95%, respectively.

In an objective theory of inference, a probability or degree of certainty is determined or measured as an observable frequency in accord with the general scientific principle of verifiability, which, for statistical inference, may be termed the

Confidence Principle:

> 'The probability of truth of a proposition is to be equated in magnitude with the frequency of truth (confidence frequency) in the most relevant generic class of propositions to which the actual proposition belongs.'

An example of such a generic class might be

$$\{\bar{x} - 2SE < \mu < \bar{x} + 2SE\}; \text{ Confidence Frequency} \doteq 95\% \ ,$$

with $2.1 < \mu < 5.3$ obtained when the observed $\bar{x}_a$ is substituted for the variable $\bar{x}$ (along with its SE).

What constitutes the most relevant generic class does not become logically determinate until a complete statement of inference is considered, comprising all possible probability levels, in other words a whole probability distribution. For then a self-evident corollary of the Relevance Principle applies, which may be termed the

Equivalence Principle:

> 'A complete inferential statement must be logically equivalent to the relevant observational evidence for it.'

Otherwise it could be argued that evidence has been lost in the logical inversion of it to inferential form.

In the rare Objective Bayesian case discussed above the Confidence, Equivalence and Likelihood principles logically imply, for the unknown parameter θ_a, the probability distribution function

$$P^{(\theta_a)}(\theta;t_a) = F^{(\theta|t_a)}(\theta) \tag{3}$$

or, to identify the particular and generic forms of proposition,

$$p = P(\theta_a < \theta_p(t_a)) = F^{(\theta|t_a)}(\theta < \theta_p(t_a));\ 0 < p < 1, \tag{4}$$

in which θ is the generic variable whose conditional, posterior distribution determines the relevant frequencies.

In the far more common non-Bayesian domain of parametric inference, an additional ordering requirement becomes logically necessary to deduce an inferential probability distribution, and this is supplied by another principle basic to most scientific reasoning, the

Ordering Principle:

> 'Meaningful ordering relations between observables and parameters are relevant evidence and, by the Equivalence Principle, should be preserved in the inferential inversion of the evidence.'

For if a scientist knows, say, that his instrument has the monotone property of producing larger measurements for larger unknowns, apart from error fluctuations, he rightly expects to infer, vice versa, that larger measurements imply larger values for the corresponding unknowns, apart from the uncertainty engendered by error fluctuations. The only way of specifying this monotone property precisely is in terms of a distribution function defined on ordered spaces, almost universally R^1 (or subspaces of R^1). (The circle is discussed briefly in Wilkinson (1977)). Thus if $t, \theta \in R^1$, and if the distribution of t is non-discrete* and

$$F(t;\theta) \downarrow \theta \quad \text{for every } t\,, \tag{5}$$

then conversely the generic relation

$$P(\theta;t) \downarrow t \quad \text{for every } \theta \tag{6}$$

must hold for an inferential probability distribution for θ given t, in accordance with the Equivalence Principle. (The opposite case $F(t;\theta) \uparrow \theta$ is covered by changing the sign of one variable.)

Considering first, therefore, the case of one unknown scalar parameter $\theta_a \in R^1$, and conditioned minimal sufficient evidence for θ_a of the form $[t_a;\{dF_a(t;\theta)\},\ t \in R^1]$, where $dF_a(t;\theta) = dF(t|z_a;\theta)$ is a conditioned sampling distribution determined by the observed values z_a of ancillary statistics z (if any), the evidence may be described as Fully Relevant if, additionally, $F_a(t;\theta) \downarrow \theta$ for every t and the family $\{dF_a(t;\theta)\}$ is boundedly complete, which ensures that no further conditioning is possible, nor any contraction of the sample space $\{t\}$, or in other words that there is no nontrivial property of $dF_a(t;\theta)$ which is independent of θ (and thus irrelevant).

Given fully relevant evidence of this form, and additionally that $dF_a(t;\theta)$ is non-discrete, then the Equivalence and Confidence principles uniquely determine the inferential probability distribution function for θ_a as

$$P^{(\theta_a)}(\theta;t_a) = 1 - F_a^{(t)}(t_a;\theta)\ , \tag{7}$$

which, in the generic form $P(\theta;t)$, is seen to preserve the monotone ordering relation between t and θ. This is the Fisherian fiducial distribution for θ_a, which may also be termed a fiducial-confidence distribution to emphasize its confidence frequency interpretation.

---

*For extension to the discrete case, see Wilkinson (1977).

To display the particular and generic forms of proposition and the nature of the logical inversion to inferential form, the definition (7) may be expressed more revealingly as follows: For $0 < p < 1$, and with $\theta_p^{-1}(\theta)$ identified as the $(1 - p)^{th}$ quantile of t given θ (which enables t_a to be replaced by t on the R.H.S.),

$$\begin{aligned} p = P(\theta_a < \theta_p(t_a)) &= F^{(t)}(\theta < \theta_p(t)) , \\ &= F^{(t)}(t > \theta_p^{-1}(\theta)) , \\ &= F^{(F_a(t;\theta))}(F_a(t;\theta) > 1 - p) . \end{aligned} \tag{8}$$

Note that the generic proposition $\theta < \theta_p(t)$ is simply the logical inversion of a proposition about the value of the distribution function $F_a(t;\theta)$, which has a known retangular distribution $R(0,1)$ independent of θ. The generic proposition thus has a known confidence frequency, as required. In the generic class of propositions $\{\theta < \theta_p(t)\}$, both θ and t are variable, θ arbitrarily so, but the confidence frequency is invariantly p for every subclass $\{\theta < \theta_p(t);\ \theta \text{ given}\}$.

The function $F_a(t;\theta)$ above, which is monotone in both its arguments and distributed independently of θ, may be termed a pivotal variate for the inference. Simpler forms of it may exist; for instance if $x \sim N(\theta,1)$, $(x - \theta) \sim N(0,1)$ is a pivotal variate which leads immediately to the inferential distribution $\theta_a \sim N(x_a,1)$. Note that a restriction on the range of θ_a may induce an incomplete fiducial distribution with some probability not assigned within the range. See Wilkinson (1977).

3.4. <u>The Calculus of Inferential Probability</u>

In both of the Bayesian and non-Bayesian cases considered, the deduced form of inferential probability distribution specifies a definitional isomorphism between a probability and corresponding confidence frequency. However, there is an important asymmetry between the cases. In the Bayesian form (4), it is the <u>subject</u> θ of the generic proposition $\theta < \theta_p(t_a)$ whose conditional posterior frequency distribution determines the confidence frequency; whereas in the fiducial case (8) it is sample variable t in the predicate $\theta_p(t)$ of the proposition $\theta < \theta_p(t)$ which provides the frequency interpretation.

This asymmetry has radical implications for the calculus of probability. But first note that the Confidence Principle itself makes no such distinction between the two cases, i.e., the objective principle of scientific verifiability is equally satisfied in both cases. That the majority of mathematical statisticians other than Fisher himself have failed to recognize this may perhaps ultimately be seen as one of the most serious errors in the history of mathematics.

In studying the nature of the isomorphism between probability and frequency it is appropriate to ask under what transformations of the parameter space is the definitional correspondence of probability and frequency maintained. In other words, if we transform $\theta \to \phi(\theta)$ does the corresponding transformation $dP(\theta) \to dP(\phi(\theta))$ preserve the confidence frequency interpretation? In the Bayesian case the answer is clearly yes for any such transformation, because of the direct form of isomorphism. Classical 'probability' theory thus determines the calculus of Bayesian inferential probability. However, in the fiducial case, the functional transformation $dP(\theta) \to dP(\phi(\theta))$ preserves a confidence frequency interpretation only if $\phi(\theta)$ is an invertible function of θ. Otherwise the confidence interpretation is lost and the resulting probability distribution therefore invalid, for if $\phi(\theta)$ is non-invertible in θ there exists no function $\phi_p(t)$ such that the generic proposition $\phi(\theta) < \phi_p(t)$ can be functionally converted to a proposition about the pivotal variate $F_a(t;\theta)$, which would otherwise determine the confidence frequency.

Thus we have the radical property of fiducial inference that non-equivalent functions (not invertibly related) of a parameter in R^1 must have non-equivalent or, as I have termed it, noncoherent inferential probability distributions (if they exist). A generalization of classical probability theory is therefore involved, and the Kolmogorov axiomatization becomes inappropriate, chiefly because the abstract, set-theoretical formulation formally ignores the ordering relations scientifically essential to the deduction of a fiducial probability distribution. Let me stress that it is the axioms which are at fault in this context, not fiducial probability itself, whose scientific interpretation is quite clear and consistent with other more familiar usages of probability.

This apparently startling property of noncoherence does in fact make a good deal of scientific sense, for the same problem of non-invertible transformation affects the notions of sufficient and relevant evidence too. Consider the simple case $x \sim N(\theta,1)$. If $[x_a,\{dF(x;\theta)\}]$ is fully relevant evidence for θ_a, it clearly is not so for θ_a^2, for if θ_a is unknown the sign of x_a is obviously irrelevant evidence regarding θ_a^2. In fact, with the necessary extension of the definition of fully relevant evidence for two parameters (see below), in this case θ_a^2 and sign (θ_a), the fully relevant evidence for θ_a^2 is $[x_a^2;\{dF(x^2;\theta^2)\}]$, in which x^2 has a non-central $\chi^2(1,\theta^2)$ distribution; and thus

$$P(\theta_a^2 < \theta^2) = F(\chi^2(1,\theta^2) > x_a^2) \ . \tag{9}$$

Noncoherence of the probability distributions for non-equivalent functions of a parameter thus flows from the non-equivalence of the relevant evidence for each.

3.5. Extensions to Two or More Scalar Parameters

This is dealt with more fully in Wilkinson (1977), and I shall simply state here a definition of fully relevant evidence in factorized form for two parameters $\theta_{1a},\theta_{2a} \in R^1$, and the probability distributions derived therefrom:

If $[t_{1a},t_{2a};\{dF(t_1,t_2;\theta_1,\theta_2)\};\ t_1,t_2 \in R^1]$ is minimal sufficient evidence for θ_{1a} and θ_{2a} jointly, and fully conditioned with respect to any relevant ancillaries, and if

$$dF(t_1,t_2;\theta_1,\theta_2) = dF(t_1;\theta_1)\cdot dF(t_2|t_1;\theta_1,\theta_2) \quad , \qquad (10)^*$$

with the properties that each factor determines fully relevant evidence (with positive monotone ordering) for θ_{1a} and $\theta_{2a}|\theta_{1a}$, respectively, when the other factor is ignored, and if additionally the factorization (10) satisfies a cross-coherence condition which in simplest cases is that $F(t_2|t_1;\theta_1,\theta_2)$ is invertibly related to θ_1 (as well as θ_2), then we have a fully relevant separation of evidence for θ_1 and $\theta_2|\theta_1$ respectively, leading to

$$P^{(\theta_{1a})}(\theta_1;t_{1a}) = 1 - F^{(t_1)}(t_{1a};\theta_1) \quad , \qquad (11)$$

$$P^{(\theta_{2a}|\theta_{1a})}(\theta_2;t_{2a},t_{1a},\theta_{1a}) = 1 - F^{(t_2|t_{1a})}(t_{2a};\theta_2,\theta_{1a},t_{1a}) \quad , \qquad (12)$$

and hence the simultaneous distribution in factorized form

$$dP(\theta_{1a},\theta_{2a}) = dP(\theta_{1a})\cdot dP(\theta_{2a}|\theta_{1a}) \quad , \qquad (13)$$

from which the <u>expected</u> fiducial distribution $d\bar{P}(\theta_{2a})$ for θ_{2a} is given by the expectation integral

$$\bar{P}(\theta_{2a}) = \int P(\theta_{2a}|\theta_{1a})dP(\theta_{1a}) \quad . \qquad (14)$$

$d\bar{P}(\theta_{2a})$ has an <u>expected</u> confidence frequency interpretation.

Note that the crosscoherence condition is a purely logical requirement, necessary for logical consistency in defining an expectation as in (14). The crosscoherence condition verifies that $P(\theta_{2a}|\theta_{1a})$ is fully dependent on θ_{1a} and not merely on some degenerate function of θ_{1a} such as θ_{1a}^2, for then $dP(\theta_{1a}^2)$ would need to be substituted for $dP(\theta_{1a})$ in the expectation integral (14). This eliminates all the earlier paradoxes of fiducial theory, other than those attributable to noncoherence.

---

*Similar conclusions flow from an alternative factorization to (10),

$$dF(t_1|t_2;\theta_1)\cdot dF(t_2;\theta_1,\theta_2) \quad .$$

3.6. Fisher's Solution for Location and Scale Parameters

This is of particular relevance for the numerical studies in Section 5. Consider a location-scale family of distributions with a continuous density function of the form

$$\frac{1}{\sigma} f\left(\frac{x - \mu}{\sigma}\right) , \tag{16}$$

and a sample $x_1, x_2, \ldots, x_n$ of independent observations (omitting the subscript $\underline{a}$) for which μ and σ are unknown. The Fisherian estimation theory for this case was first given in Fisher (1934), see also Fisher (1956). Here the relevant ancillary statistics for conditioning specify the complexion or shape of the sample, and may be conveniently expressed in the form

$$\frac{x_i - m}{s}, \quad i = 1,2,\ldots,n , \tag{17}$$

where m and s are any convenient estimators of location and scale derived from the sample, with s independent of location. Only $n - 2$ of the ancillaries are functionally independent. Conditioning the sampling distribution of (m,s) with respect to the ancillaries (17) reduces the evidence to fully relevant form.

Logical inversion of this fully relevant evidence produces a simultaneous fiducial-confidence distribution for μ and σ, conditioned by the sample complexion, which was first given in Fisher (1948). See Fisher (1956) for derivation. In simplest form the distribution is specified as

$$dP(\mu,\sigma) \propto \prod_{i=1}^{n} \left\{\frac{1}{\sigma} f\left(\frac{x_i - \mu}{\sigma}\right)\right\} d\mu \frac{d\sigma}{\sigma} , \tag{18}$$

which, because of the group-invariant properties of the family (16), is simply the likelihood of the sample multiplied by $d\mu d\sigma/\sigma$. Integrating with respect to σ gives an expected fiducial-confidence distribution $d\bar{P}(\mu)$ for μ. See (22), Section 5 for an illustration.

In view of the formal resemblance of (18) to a Bayesian posterior distribution derived from an assumed prior $d\mu d\sigma/\sigma$, it is important to stress that the objective confidence frequency interpretation of $dP(\mu,\sigma)$ and $d\bar{P}(\mu)$ is quite different from a Bayesian one, and is independent of any such assumption.

The solution (18) embodies implicitly just the right degree of conditioning on sample complexion, depending on f. In the Normal case the solution turns out to depend only on the the jointly sufficient statistics $\bar{x}$, s, leading to the familiar results $\mu \sim \bar{x} + st_{n-1}$, $\sigma^2 \sim (n-1)s^2\chi^{-2}_{n-1}$. Similarly, if f is a uniform density, the solution depends only on the range and midrange of the sample.

3.7. Extension to Approximate Inference Theory

Exact fiducial-confidence distributions of the kind described exist for a wide range of important cases but, more generally, some degree of indeterminacy arises because there is a degree of inherent confounding of relevant with irrelevant evidence, or of the relevant evidence for two or more parameters. However, the indeterminacy in separating the different kinds of evidence disappears asymptotically, so that development of the right kind of conditional asymptotic theory should provide guidance on adequate approximate separations of the evidence in finite samples.

Of more relevance to robustness studies is that the sampling distribution of the data is not usually known exactly (in parametric form). But here the right parametric approach is clearly to formulate a larger family of sampling distributions, introducing additional shape parameters for instance, so that the larger family is sufficiently representative to contain an adequate approximation to the underlying true distribution for inferential purposes. However, I do not believe that an endless proliferation of sampling distributions will prove to be necessary. I think future robustness studies will show that only certain salient features of distribution need be allowed for in the analysis, with other more irregular features having negligible effect. Likewise, ancillary conditioning will prove to be dependent only on some correspondingly salient features of the sample - a conditional form of robustness, of which a good illustration appears in the numerical studies in Section 5.

In seeking a formal extension of a theory of statistical inference to approximate inferences, one may note first a peculiar feature of an uncertain inference, in contrast with a statement of definite fact, namely that its very uncertainty renders a degree of approximation automatically tolerable. We may express this in the form of an

Uncertainty Principle (First)*:

> 'The degree of uncertainty about the specified uncertainty in an uncertain inference need be no less than commensurate with the specified uncertainty itself.'

Secondly, an assumption of continuity plays an important part in extending the domain of application of a theory. In statistical inference, continuity enters as a basic empirical requirement:

Continuity Principle:

> 'Small perturbations of the evidence should produce corresponding small changes in the inferences therefrom.'

Note that the intention is not necessarily to exclude any discontinuous effect of a small change, but to force a thoroughly valid explanation of any that does arise.

4. HARTIGAN CONFIDENCE DISTRIBUTIONS

Hartigan (1969) has derived an important generalization in the context of confidence intervals of Fisher's (1935) Randomization test (described in the Introduction). He shows that, given a set of independent sample variables $x_1, x_2, \ldots, x_n$, and assuming no more than that their sampling distributions are continuous and symmetric about a common location value μ, a multiplicity of distribution-free confidence distributions (my term), or typical sets (Hartigan's term), can be derived for the parameter μ, depending on the choice of a location estimator.

---

*There are possibly other uncertainty principles, one at least partly expressed in G. E. P. Box's concept of a 'parsimonious' model.

Hartigan defines a typical set of N values for μ as an ordered set such that the $N + 1$ intervals between successive values including $\pm\infty$ are exact confidence intervals for μ, each covering μ with equal probability $1/(N + 1)$. Regarded alternatively, a histogram constructed by grouping the values of a typical set forms a confidence distribution for μ, whose only inexactness is that engendered by the grouping.

Given the sample $x_1, x_2, \ldots, x_n$ as above, consider the complete set of all $N = 2^n - 1$ possible subsamples of it, including the sample itself. Then the N arithmetic means of all the samples in this set form a typical set of values, or, in my terminology, a confidence distribution for μ in the sense described. Hartigan shows this to be precisely the confidence analogue of Fisher's Randomization test.

He then proceeds to the more general result, namely, that applying any of a wide class of symmetrically distributed estimators to the complete set of samples as above will produce a corresponding confidence distribution (or typical set), appropriate to that estimator. Hartigan also defines balanced subsets of the complete set of samples, and his theorems on typical sets refer more generally to balanced subsets. However, we shall not be concerned with balanced subsets here.

The main condition on an estimator for forming a Hartigan confidence distribution for μ is that it be functionally the solution of an estimating equation

$$\Sigma\, e(x_i, \mu) = 0 \ , \tag{19}$$

with summation over the values x_i in the relevant sample, where $e(x,\mu)$ is an 'estimating function' which is continuously and symmetrically distributed about zero. The requirement for $e(x,\mu)$ to be an estimating function is that it be continuous and decreasing as a function of μ, crossing the abscissa for every x, and that $E\{e(x,\mu)\} = 0$. The functional constraints on an estimating function ensure that the estimating equation (19) always has a unique solution for μ. However, a weaker sufficient condition for present purposes would be that (19) should give unique solutions for the actual set of N samples being considered, with $\Sigma\, e(x_i,\mu)$ locally decreasing at the solution in each case.

Maximum-likelihood estimators, in particular, applied to symmetrically distributed data as above, satisfy the requirements for producing a Hartigan confidence distribution if the likelihood function of any sample is log-concave in μ. Here $e(x,\mu)$ represents Fisher's Score Function for an observation, and (19) the Equation of Maximum Likelihood, equating the score function for the sample to zero. In Section 5 the maximum-likelihood estimators of location for a family of exponential-power distributions will be applied to generate a corresponding family of Hartigan confidence distributions for various samples.

It is important to stress, when it comes to comparing Hartigan distributions with the corresponding Fisher conditional distributions, that the confidence frequency interpretation of a Hartigan distribution is based on marginal frequencies.

5. NUMERICAL COMPARISONS

The main objective of the numerical studies described here is to compare, for various samples, and for various assumed sampling distributions, the Fisher conditional fiducial-confidence distribution for location with the corresponding Hartigan marginal, distribution-free confidence distribution derived by applying the corresponding maximum-likelihood estimator to the sample and to every subsample of it ($2^n - 1$ estimations in all). The family of sampling distributions considered is the exponential-power family with density function of the symmetric form

$$f(x) \propto \frac{1}{\sigma} \exp\left\{-\frac{1}{2}\left|\frac{x-\mu}{\sigma}\right|^q\right\}, \quad q \geq 1\ , \tag{20}$$

with location-scale parameters μ, σ and a shape parameter q.

For any sample $x_1, x_2, \ldots, x_n$ the log-likelihood function is strictly concave w.r.t. μ if $q > 1$. When $q = 1$ it is semi-concave, comprising linear segments between the abscissae $\mu = x_1, x_2, \ldots, x_n$. The maximum-likelihood estimate of location μ is obtained by minimizing

$$M(\mu) = \sum_i^n |x_i - \mu|^q ; \tag{21}$$

and the Fisher fiducial-confidence distribution for μ has density function

$$p(\mu) \propto M^{-n/q} . \tag{22}$$

With $q = 2/(1 + B)$ as in Box and Tiao (1962), $q \geq 1$ corresponds to $-1 < B \leq 1$, and the special cases of the Normal, double exponential and rectangular distributions are specified by $B = 0, 1$ and (in the limit) -1, with corresponding m.ℓ. estimators the mean, median and midrange, respectively.

For continuity of discussion with Fisher (1935, Chapter III), and Box and Tiao (1962), the same Darwin data have been included in the study. They comprise the 15 values ordered, in 1/8-inch units of height difference,

-67 -48 6 8 14 16 23 24 28 29 41 49 56 60 75 .

Note the two negative outliers, which are significantly in deviation from a Normal distribution. However, the most important conclusions flow from a set of artificial samples, each comprising 15 quantiles at equi-spaced probability levels

$$P = 1/30, 3/30, \ldots, 29/30 , \tag{23}$$

of an exponential-power distribution with $B = -0.4, 0, 0.4, 0.8, 1.2$ respectively, which give an increasing degree of stretch in the tails.

All samples have been standardized for comparison as deviations from the sample mean in units of Adjusted Standard Error (ASE) of the mean defined by $ASE^2 = Sx^2/(n - 3)/n$. The divisor $n - 3$ in place of $n - 1$ assigns unit variance to the corresponding adjusted t-statistic MEAN/ASE. The standardized values are listed in Table 1, together with the sample skewness (G3) and kurtosis (G4) coefficients. The comparative configurations of the standardized samples are shown in Figure 1.

TABLE 1

(a) Darwin sample and five quantile samples generated from an exponential-power family of distributions with B values as shown. Quantiles correspond to equi-spaced probabilities as shown. Samples standardized as deviations from sample mean in ASE units (see text).

	Darwin sample	Sample B =	Generated quantile samples -0.4	0.0	0.4	0.8	1.2
1	-8.354	30P= 1	-6.207	-6.629	-6.976	-7.25	-7.329
2	-6.549	3	-4.701	-4.630	-4.544	-4.452	-4.464
3	-1.419	5	-3.721	-3.497	-3.286	-3.098	-3.037
4	-1.229	7	-2.897	-2.631	-2.391	-2.183	-2.091
5	-0.659	9	-2.146	-1.895	-1.673	-1.484	-1.389
6	-0.469	11	-1.422	-1.232	-1.058	-0.914	-0.834
7	0.196	13	-0.710	-0.607	-0.509	-0.429	-0.381
8	0.291	15	0.000	0.000	0.000	0.000	0.000
9	0.671	17	0.710	0.607	0.509	0.429	0.381
10	0.766	19	1.422	1.232	1.058	0.914	0.834
11	1.906	21	2.146	1.895	1.673	1.484	1.389
12	2.666	23	2.897	2.631	2.391	2.183	2.091
13	3.331	25	3.721	3.497	3.286	3.098	3.037
14	3.711	27	4.701	4.630	4.544	4.452	4.464
15	5.136	29	6.207	6.629	6.976	7.252	7.329
Skewness and Kurtosis Coefficients:							
G3	-1.106		0	0	0	0	0
G4	1.408		-0.764	-0.309	0.156	0.591	0.738

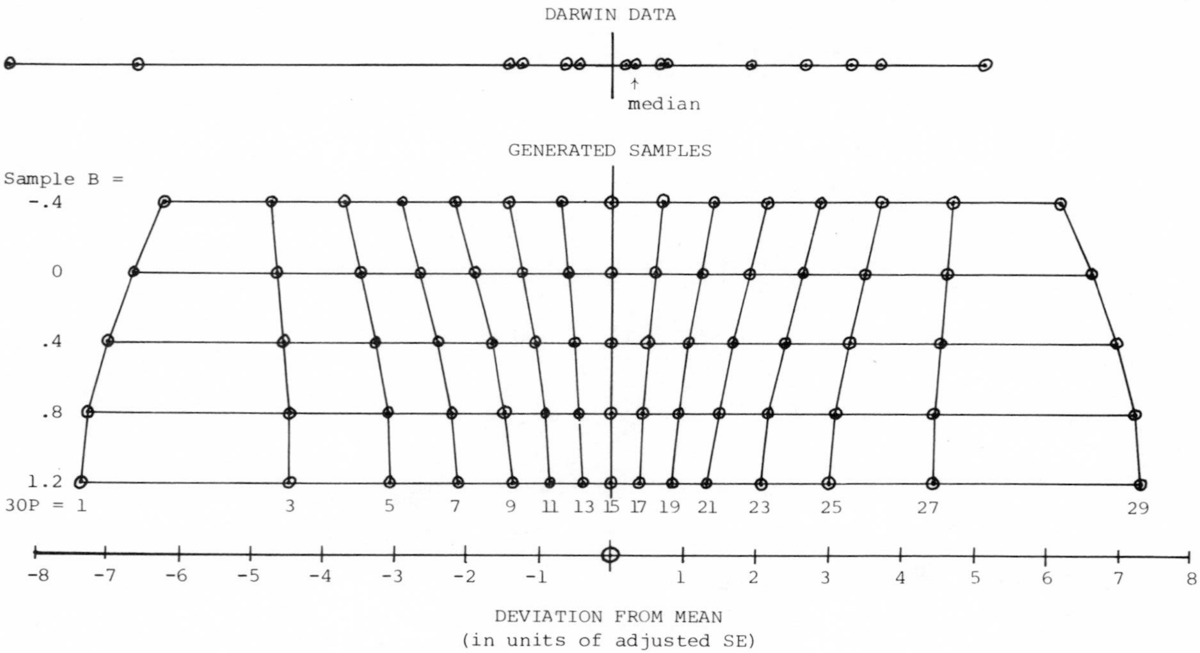

FIGURE 1. Darwin sample and artificial quantile samples generated from an exponential-power family of distributions with B values as shown. Quantiles correspond to equi-spaced probabilities as shown.

Fisher and Hartigan confidence distributions have been calculated from each sample for a range of assumed values of the shape parameter B which, for the corresponding Hartigan distribution, determines the relevant maximum likelihood estimator. The B values ranged in steps of 0.2 from -1 (-0.2 in Figure 3) to +1, with supplementary calculations at B = 0.84, 0.88, 0.92, 0.98 to produce more detail of the Hartigan distribution behaviour as B approaches unity. Hartigan distributions for the special case B = 1 are omitted, since in a comparative sense they become partly indeterminate, due to the relevant maximum likelihood estimates being uniquely defined as medians only for odd-sized samples. (For an even-sized sample the maximum likelihood solution comprises the whole interval between the two middle values. Defining the median conventionally as the mid-value between these limits does not satisfy the requirements of the theory.)

Each Hartigan distribution involved 32,767 m.ℓ. determinations derived (except for means and midranges) by a 3-stage search of the function $M(\mu)$ for each subsample, at intervals of 0.4, 0.2, 0.1 ASE units respectively. However, the net computing cost for each distribution was less than a dollar.

The results are shown in Figure 2 (artificial samples) and Figure 3 (Darwin sample). Each ordinate of a curve represents, for a particular sample and confidence level (90, 95 or 99%), the deviation of a confidence interval endpoint from that for the Normal case (a quantile of the distribution of t_{14}, adjusted to unit variance); and each curve shows how such a confidence interval deviation depends on the B value assumed for the calculation. Only upper endpoint deviations are shown for the symmetric artificial samples, but both upper and lower for the skew Darwin data, with reversed signs for lower endpoint deviations so that points above or below the abscissa represent, respectively, deviation to longer or shorter than Normal-case intervals. Superimposed curves in Figure 2 contrast the effects of different samples. In Figure 3, they contrast upper with lower, Fisher with Hartigan.

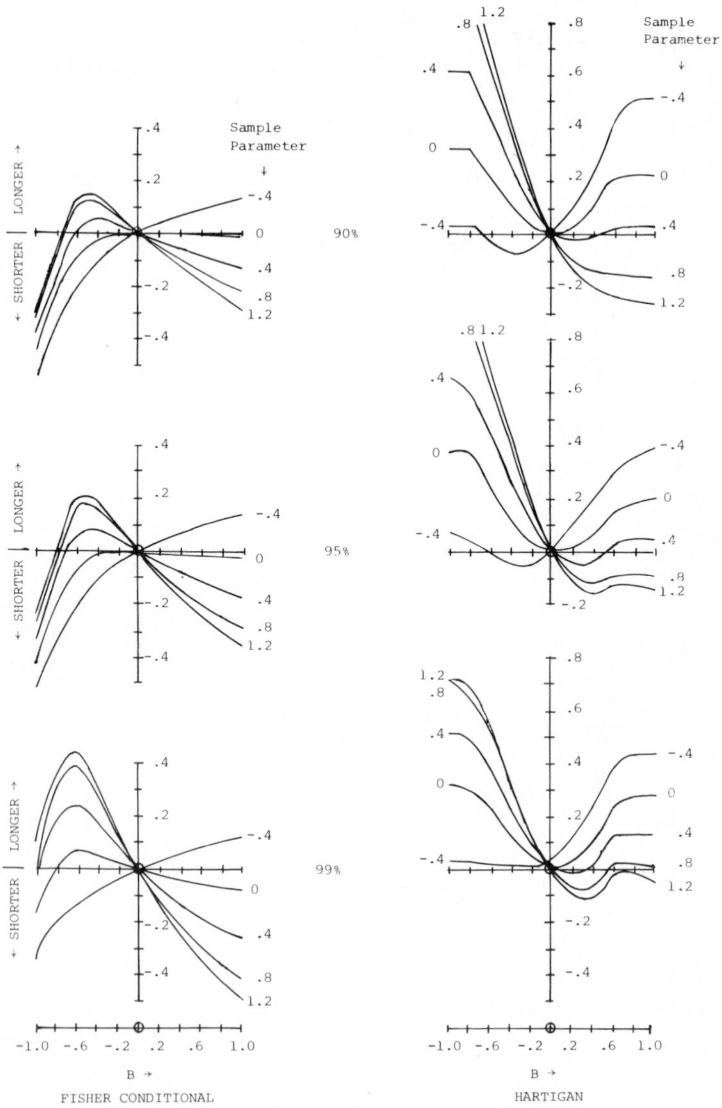

FIGURE 2. Deviations (in ASE units) of upper confidence-interval endpoints from Normal-case (t_{14}). Generated samples.

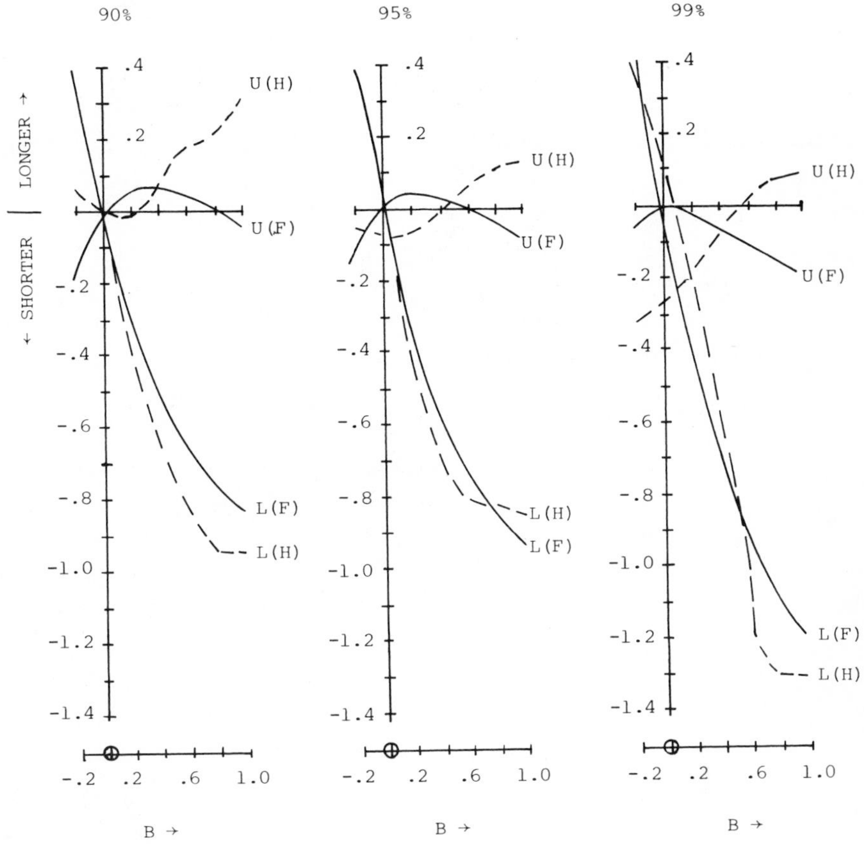

FIGURE 3. Deviations (in ASE units) of confidence interval endpoints from Normal-case (t_{14}). Darwin data. U, L: Upper, lower endpoints; F, H: Fisher, Hartigan distributions. Deviations reversed in sign for lower endpoints.

In interpreting the curves note that each increment of 0.1 in an endpoint deviation reduces the corresponding one-tail probability by 20-25%, for the 90-99% range of confidence levels.

The curves for the Fisher distributions are of course concurrent by definition at the origin $B = 0$, $DEV = 0$, corresponding to a classical Normal analysis. The remarkable empirical result, however, is that the Hartigan curves are also extremely close to concurrence at this same origin, specially for the artificial samples, though less so for the Darwin data at the 99% level in particular. This is a striking central-limit effect for Hartigan distributions, as the estimator which generates them tends in form to the arithmetic mean; and somewhat orthogonal to the effect of increasing sample size as in classical central-limit theory.

A second striking central-limit effect of a different character again is to be seen in the curves for the Fisher conditional distributions in Figure 2. All the conditional curves show just the kind of sensitivity to the underlying assumption concerning distribution (the B value) that a good data-analyst might expect, but as the degree of departure of the sample from apparent Normality decreases, so does the sensitivity to the distributional assumption, so that in the limit, for a sample that looks completely Normal (sample $B = 0$), the final confidence analysis is almost independent of the distribution assumed and agrees with the Normal analysis over the whole range of long-tailed (leptokurtic) distributions ($B = 0$ to 1), and to a lesser extent over a range of moderately platykurtic distributions (down to say $B \approx -0.5$). (Had a sample comprising 15 Normal Rankits been included in the study, the agreement with Normal analysis would have been even closer.)

Clearly there is a whole new domain of conditional central-limit theory to be developed here, as mentioned in the Introduction, in which it is the appearance of the sample that counts. Figure 3 for the skew Darwin data also suggests nearly independent effects of the appearance of the left- and right-hand sides of the sample.

Differently expressed, we have here a new, conditional parametric concept of robustness which is the obverse of Criterion Robustness, for here it is the method of analysis which varies, leading nevertheless to the same result. The primary influence of Conditional Robustness as a potentially attainable property of the analysis will be to place more emphasis on transformation to canonical forms in which the latent robustness is realized; of which transformation of a sample to closer apparent Normality is an important example.

Returning to the conditional fiducial analysis illustrated in Figure 2 and Figure 3: If in fact the shape parameter B is unknown, the Fisherian theory requires that this additional element of uncertainty be incorporated (at least at the final stage of analysis) by averaging the conditional confidence distributions given B with respect to a fiducial distribution for B. In analytic terms this is an unsolved problem at present. Clearly the maximum likelihood estimate of B from the sample is relevant, but its appropriately conditioned sampling distribution will depend on further ancillary statistics for shape, of which a goodness-of-fit statistic for the assumed family of distributions would be the most important; for clearly the evidence about a shape parameter of the family will be stronger with a good fit than a bad one. (D. V. Hinkley (pers. comm.) is also aware of this point.)

Fiducial averaging with respect to possible values of the shape parameter B above will tend to pull the resultant fiducial distribution for location closer to the Normal form. If indeed shorter-than-Normal-tailed distributions were excluded on *a priori* grounds, the fiducial analysis would assign a definite positive weight to the Normal form in the averaging. See Box and Tiao (1962) for the corresponding Bayesian viewpoint.

We come finally to a general comparison of the conditional Fisher distributions and their marginal, distribution-free Hartigan counterparts, first for the artificial samples in Figure 2. The degree of agreement at $B = 0$ extends over a moderate range of B values, say $-0.3 < B < +0.3$, but outside this range the Hartigan confidence distributions become, for this set of samples, considerably more conservative than their Fisherian counterparts, and often more conservative to a marked degree than the Normal form (t_{14}).

The dominant reason for this is not hard to find. In considering estimates from all possible subsamples one is including in the enumeration an appreciable fraction of subsamples with shape quite unlike that of the whole sample, on which the Fisherian analysis is conditioned. These 'atypical' subsamples produce potentially serious distortions of the inferential distribution and leave little doubt, pragmatically, of the need for conditioning which Fisher so often stressed. In the case of the Darwin data, Figure 3, the story is similar for the upper tail, but there appears to be reasonably good agreement for the lower interval endpoints, which are extremely sensitive to the assumed B value because of the two extreme outliers. I think the apparent agreement here is due to the fact nearly all the subsamples that determine the lower confidence tail include at least one of the bad outliers and thus have a similar shape to the whole sample in this dominating respect.

6. CONCLUSIONS

We have seen in Section 3 that six basic principles of statistical inference (Relevance, Equivalence, Confidence, Sufficiency, Conditioning and Ordering) lead logically to Fisherian fiducial-confidence distributions as the appropriate form of inferential probability distribution in a wide domain of parametric inference; and that the same principles lead also to a generalization of classical probability theory for that domain, in which nonequivalent functions on a parameter space have nonequivalent (noncoherent) probability distributions, because of the nonequivalence of the relevant evidence for each kind of inference. One further consequence of note is that, in this generalized calculus of probability, Bayes' Theorem is no longer generally true as a theorem for modifying an inferential probability distribution in the light of further observational evidence. This raises serious doubts about the validity of axiomatic Bayesian theories.

Perhaps the main implication for robustness research is that the operational role of the many other mathematical principles which have been introduced needs to be critically re-examined. Though some of these have an asymptotic validity for large samples, they are, for finite samples, in fundamental conflict with the basic principles of inference described here, which allow no room logically for any other such principles to operate. Indeed, since violation of the basic principles of Fisherian theory makes it intrinsically possible to exaggerate artificially the real strength of evidence in the data, the substitution of other mathematical optimization principles ('shortest' intervals and so forth) is very likely to achieve just such an effect, and therefore needs the closest scrutiny.

We have also seen, from the numerical studies in Section 5, how the Fisherian analysis points to a new, Conditional Robustness property, arising as a Conditional central limit effect of an increasing approach of the sample shape to apparent Normality, irrespective of sample size; and indeed it is likely that this conditional approach will render obsolete many of the standard results in central limit theory. It must be stressed, however, that we should not now introduce yet another principle, of optimizing Conditional Robustness (if considered as a new criterion of choice between intrinsically different methods of analysis), for that again would be in conflict with the basic principles of statistical inference, which allow no such choice. Instead we may properly seek, by transformations etc., to reveal and utilize any robustness properties that are latent in the appropriate conditional analysis for the data, thereby simplifying the ultimate form of that analysis, particularly at a practical level.

REFERENCES

Box, G. E. P. (1953). Non-normality and tests on variances. Biometrika, 40, 318-335.

Box, G. E. P. and Tiao, G. C. (1962). A further look at robustness via Bayes's theorem. Biometrika, 49, 419-432.

Cox, D. R. (1958). Some problems connected with statistical inference. Ann. Math. Statist., 29, 357-372.

Darwin, Charles (1876). The Effects of Cross and Self Fertilization in the Vegetable Kingdom. John Murray, London.

Diananda, P. H. (1949). Note on some properties of maximum likelihood estimates. Proc. Camb. Phil. Soc., 45, 536-544.

Fisher, R. A. (1922). On the mathematical foundations of theoretical statistics. Phil. Trans., A., 222, 309-368.

____________ (1934). Two new properties of mathematical likelihood. Proc. Roy. Soc., A, 144, 285-307.

____________ (1935). The Design of Experiments. Oliver and Boyd, Edinburgh. (8th edition, 1966.)

____________ (1948). Conclusions fiduciaries. Ann. Inst. Henri. Poincaré, 10, 191-213.

____________ (1956). Statistical Methods and Scientific Inference. Oliver and Boyd, Edinburgh. (3rd edition, Hafner Press, New York, 1973.)

Hartigan, J. A. (1969). Using subsample values as typical values. J. Amer. Statist. Ass., 64, 1303-1317.

Tukey, J. W. (1977). Exploratory Data Analysis. Addison Wesley, Reading, Mass.

Wilkinson, G. N. (1977). On resolving the controversy in statistical inference. J. Roy. Statist. Soc., B, 39, 119-171.

Sponsored by the United States Army under Contract No. DAAG29-75-C-0024. The author is also indebted to Cristobal Vargas for computing assistance.

Mathematics Research Center
University of Wisconsin-Madison
Madison, WI 53706

and

Genetics Department
University of Adelaide
South Australia

(Present address)

AUTHOR INDEX

SUBJECT INDEX